T0176438

IP FOR 4G

IP FOR 4G

Dave Wisely
BT, UK

A John Wiley and Sons, Ltd, Publication

Library of Congress Cataloging-in-Publication Data

Wisely, David.
  IP for 4G / David Wisely.
     p. cm.
  Includes bibliographical references and index.
  ISBN 978-0-470-51016-2 (cloth)
 1. Wireless Internet. 2. Global system for mobile communications.
3. Mobile communication systems. I. Title.
  TK5103.4885.W5732 2009
  621.3845'6–dc22   2008037703

A catalogue record for this book is available from the British Library.

ISBN 9780470510162 (H/B)

Typeset by 10/12 pt Times Roman by Thomson Digital, Noida, India
Printed in Great Britain by CPI Antony Rowe, Chippenham, Wiltshire

Table of Contents

About the Author

Dr. Wisely has worked for BT for 20 years in the fields of networks and mobility research. He pioneered optical wireless links in the early 1990s constructing a 4 km, 1500 nm system using optical amplifiers.

Dave has worked in the field of mobility for the past 10 years, looking firstly at wireless ATM and HIPERLAN systems and more latterly into the combination of cellular mobile and WLAN systems.

Dave was one of the pioneers of an all-IP solution for future developments of 3G. He also acted as technical manger for the influential EU IST BRAIN/MIND EU IST project which did much to push forward with all-IP network concepts.

He has contributed over 100 papers to journals and conferences, published one previous book and contributed chapters to over a dozen others. Dave was also the co-editor of the BTTJ special edition of mobility. He is currently in charge of all convergence research and development at BT and has responsibility for BT's twenty-first century network's convergence research programme.

Dave is married with two children aged 11 and 8 and lives in rural Suffolk. His hobbies include: cricket, tram-spotting and complaining about the transport system in the UK.

Preface

This books attempts to demystify the black art of 4th Generation wireless technology that will underpin human mobility and continue to transform the way we interact.

Dave's personality oozes out from between the pages of what is a technically complex and potentially very dry read adding a fresh, real quality seldom found in technical reference titles. The underlying feeling the reader will come away with is that the good Dr Wisely would – if he could – quickly engage you in a rewarding and entertaining conversation on the subject ... peppered with a rollercoaster of stories, observations and industry insider tit-bits.

The coverage is ideal for a degree level student taking a wireless module – looking for some palatable detail for example. However, this is not to suggest that the book is a simplistic overview or lacking the technical depth for the professional reader or researcher – far from it! Some sections will leave you reeling and desperate for the headache tablets (his words) and just following a few of the numerous references will confirm that it was well worth the cover price for the distillation and skilful interpretation that only a world expert at the top of his game could provide.

Reference to Dave's family, work and light hearted jibes at the industry (especially the standards fraternity) lighten the load and squarely position the desire and enthusiasm the author has to communicate the ongoing impact of wireless technology.

Although you may choose to read it from cover to cover the book feels more like a grid that will get you from A to B than a guided tour. Some limited and carefully positioned duplication allows you to make the journey from question to answer several different ways depending on your requirements and con-straints ... overview, detail, industry perspective etc.

With the author employed by a leading communications company with a considerable commercial interest in the Business Case for 4G and an active Public Relations clamp it comes as a surprise that we are exposed to some of the more interesting industry debates. As with all published material this topical edge will fade leaving a rich resource with the time stamp of a period of intense technical innovation.

Richard Dennis CEng, MBCS, CITP
Head of BT's Mobility Research Centre

Acknowledgements

I must also extend a note of thanks to: Phil "the mallet" Eardley and Richard "I'm Offski" Dennis for reading and much improving the entire manuscript; to Richard Gedge and Maurice Gifford for much improving specific chapters; to many other colleagues for useful conversations and for clearing up detailed technical issues; to Jax for proofreading; to Sarah, Birgit and Sally from John Wiley & Sons, Ltd for badgering me into finishing the project; to BT for the opportunity to write the book; to the staff at Tesco (Martlesham and Copdock branches), Orwell Crossing café and the Bus Café on the A12 for keeping me going (and in one case finding my laptop).

List of Abbreviations

3GPP	3rd Generation Partnership Project
3GPP2	3rd Generation Partnership Project 2
AAA	Authentication, Authorization and Accounting
AAL	ATM Adaptation Layer
AC	Access Controller
ACR	Access Control Router
ACK	Acknowledgement
AD	Administrative Domain
AES	Advanced Encryption Standard
AH	Authentication Header
AIFSD	Arbitration InterFrame Space Duration
AN	Access Network
ANG	Access Network Gateway
ANR	Access Network Router
AP	Access Point
API	Application Programming Interface
AR	Access Router
A-RACS	Access Resource and Admission Control Subsystem
ARP	Address Resolution Protocol
ARPU	Average Revenue Per User
ARQ	Automatic Repeat Request
ARPANET	Advanced Research Projects Agency Network
AS	Application Server
ASN	Access Service Network
ASP	Application Service Provider
ATM	Asynchronous Transfer Mode
B-ISDN	Broadband ISDN
B-SMS	Broadcast Short Message Service
BCH	Broadcast Channel
BCCH	Broadcast Control Channel (GSM)
BE	Best Effort
BER	Bite Error Rate

BGCF	Breakout Gateway Control Function
BGP	Border Gateway Protocol
BMC	Broadcast/Multicast Control
BM-SC	Broadcast Multicast Service Centre
BPSK	Binary Phase Shift Keying
BRAIN	Broadband Radio Access for IP based Networks
BS	Base Station
BSC	Base Station Controller
BSS	Base Station System (2G and 3G) and Basic Service Set (WLAN)
BT	British Telecommunications plc
BTS	Base Transceiver Station
CAMEL	Customised Application for Mobile network Enhanced Logic
CAP	CAMEL Application Part
CAPWAP	Control And Provisioning of Wireless Access Points
CC	Call Control
CCCH	Common Control Channel
CCCF	Call Continuity Control Function
CDMA	Code Division Multiple Access
CDPD	Cellular Digital Packet Data
CH	Correspondent Host
CID	Connection IDentifier
CN	Correspondent Node
CoA	Care-of Address
COPS	Common Open Policy Service
CPS	Common Part Sub-Layer
CS	Circuit Switched and Convergence Sublayer (WiMAX)
CSCF	Call Server Control Function
CSMA	Carrier Sense Multiple Access
CSMA/CD	Carrier Sense Multiple Access with Collision Detection
CSN	Connectivity Service Network
CTCH	Common Traffic Channel
CTS	Clear To Send
DAB	Digital Audio Broadcast
DCCH	Dedicated Control Channel
DCF	Distributed Control Function
DECT	Digital Enhanced Cordless Telecommunications
DCH	Dedicated Channel
DES	Data Encryption Standard
DFS	Dynamic Frequency Selection
DHCP	Dynamic Host Configuration Protocol
DiffServ	Differentiated Services
DIFS	DCF InterFRame Space
DNS	Domain Name System

DoS   Denial of Service
DPDCH   Dedicated Physical Data Channel
DPCCH   Dedicated Physical Control Channel
DS   Differentiated Services
DSCP   Differentiated Services CodePoint
DSSS   Direct Sequence Spread Spectrum
DTCH   Dedicated Traffic Channel
DVB-H   Digital Video Broadcast - Handheld

E1   2Mbit/s TDM circuit connection
E-AGCH   E-DCH Absolute Grant Channel
EAP   Extensible Authentication Protocol
EC   European Commission
E-DCCH   E-DCH Dedicated Control Channel
E-DCH   Enhanced Dedicated Channel
EDCF   Enhanced Distributed Control Function
EDGE   Enhanced Data rates for GSM Evolution
E-DPDCH   Enhanced Dedicated Physical Data Channel
ENUM   (From) tElephone NUMber mapping
E-RGCH   E-DCH Relative Grant Channel
ESP   Encapsulating Security Payload
ETSI   European Telecommunications Standards Institute
EU   European Union
EVDO   EVolution Data Optimised
EVDV   EVolution Data and Voice

FA   Foreign Agent
FACH   Forward Access Channel
FACCH   Fast Associated Control Channel (GSM)
FBBS   Fast Base Station Switching
FCA   Fixed Channel Allocation
FCC   US Federal Communications Commission
FCH   Frame Control Header
FDD   Frequency Division Duplex
FDM   Frequency Division Multiplexing
FDMA   Frequency Division Multiple Access
F-DPCH   Fractional Dedicated Physical Channel
FEC   Forward Error Correction
FIFO   First In First Out
FPLMTS   Future Public Land Mobile Telecommunication Systems
FM   Frequency Modulation
FT   Fourier Transform
FTP   File Transfer Protocol

GERAN   GSM EDGE Radio Access Network
GFA   Gateway Foreign Agent

GGSN	Gateway GPRS Support Node
GMSK	Gaussian Mean Shift Keying
GPRS	General Packet Radio Service
GRE	Generic Routing Encapsulation
GSM	Global System for Mobile Communications (ETSI)
GTP	GPRS Tunnelling Protocol
HA	Home Agent
HACK	Handover Acknowledgement
HARQ	Hybrid Automatic Repeat Request
HCCA	HFC Controlled Channel Access
HCF	Hybrid Control Function
HI	Handover Initiate
HLR	Home Location Register
HMIP	Hierarchical Mobile IP
HSCSD	High Speed Circuit Switched Data
HS-DPCCH	High Speed uplink Dedicated Physical Control Channel
HS-DSCH	High Speed Downlink Shared Channel
HSPA	High Speed Packet Access
HSDPA	High Speed Download Packet Access
HS-SCCH	High Speed Shared Control Channel
HSUPA	High Speed Uplink Packet Access
HSS	Home Subscription Server
HTTP	Hyper Text Transfer Protocol
IAB	Internet Architecture Board
IAM	Initial Address Message
IANA	Internet Assigned Numbers Authority
ICAN	IP Connectivity Access Network
ICMP	Internet Control Message Protocol
I- CSCF	Interrogating Call Server Control Function
ID	Identity
i-DEN	Integrated Digital Enhanced Network
IEEE	Institute of Electrical and Electronics Engineers
IESG	Internet Engineering Steering Group
IETF	Internet Engineering Task Force
IFS	InterFrame Spacing
IFT	Inverse Fourier Transform
IM	Instant Messaging
IMEI	International Mobile station Equipment Identifier
IMS	Internet Multimedia Subsystem
IM-SSF	IP Multimedia Service Switching Function
IMSI	International Mobile Subscriber Identity
IN	Intelligent Network
INAP	Intelligent Network Application Protocol
IntServ	Integrated Services

IP	Internet Protocol
IPR	Intellectual Property Rights
IPsec	IP Security
IS	Information Service
IPTV	IP Television
IRTF	Internet Research Task Force
ISDN	Integrated Services Digital Network
ISIM	IP Multimedia Services Identity Module
ISM	Industrial Scientific and Medical (band)
ISP	Internet Service Provider
IST	Information Society Technologies
ISUP	Integrated Services User Part
ITU	International Telecommunication Union
IV	Initialization Vector
JPEG	Joint Photographic Experts Group
LAN	Local Area Network
LAP	Link Access Protocol
LLC	Logical Link Control
LTE	Long Term Evolution
MAC	Medium Access Control
MAP	Mobile Application Part
MBMS	Multimedia Broadcast and Multicast Services
MD5	Message Digest 5
MDHO	Macro Diversity Hand Over
ME	Mobile Equipment
MEGACO	Media Gateway Control
MG/MGW	Media Gateway
MGCF	Media Gateway Control Function
MGC	Media Gateway Controller
MH	Mobile Host
MIH	Media Independent Handover
MIMO	Multiple IN Multiple Out
MIP	Mobile IP
MSISDN	Mobile Station International Subscriber Directory Number
MM	Mobility Management
MMS	Multimedia Messaging Service
MMSC	MultiMedia Session Continuity
MMUSIC	Multiparty Multimedia Session Control
MN	Mobile Node
MNO	Mobile Network Operator
MO	Mobile Originated
MRF	Media Resource Function

MS	Mobile Station
MSAN	Multi Service Access Node
MSC	Mobile Switching Centre
MT	Mobile Terminated
MTU	Maximum Transmission Unit
NACK	Negative Acknowledgement
NAI	Network Access Identifier
NAP	Network Access Provider
NAT	Network Address Translation
NFC	Near Field Communication
NGN	Next Generation Network
NIU	Network Interface Unit
N-RTP	Non Real Time Polling
NSP	Network Service provider
OFDM	Orthogonal Frequency Division Multiplexing
OFDMA	Orthogonal Frequency Division Multiple Access
OMA	Open Mobile Alliance
OS	Operating System
OSI	Open System Interconnection
PCCH	Paging Control CHannel
PCH	Paging CHannel
PCM	Pulse Code Modulation
PCMCIA	Personal Computer Memory Card International Association
PCN	Packet Core Network (cdma2000)
P- CSCF	Proxy Call Server Control Function
PDC	Personal Digital Cellular
PDF	Policy Decision Function
PDCP	Packet Data Convergence Protocol
PDP	Packet Data Protocol
PDSN	Packet Data Service Node (cdma2000)
PDU	Protocol Data Unit
PEP	Policy Enforcement Point
PHY	Physical Layer
PIG	PSTN Internet Gateway
PIN	Personal Identification Number
PKM	Privacy Key Management
PMM	Packet Mobility Management
PoC	Push to talk Over Cellular
POTS	Plain Old Telephony Service
PPP	Point-to-Point Protocol
PSP	Play Station Portable
PSTN	Public Switched Telephone Network

PUSC	Partial Use of Sub Channels
PVC	Permanent Virtual Circuit (in ATM)
QAM	Quadrature Amplitude Modulation
QoS	Quality of Service
QPSK	Quadrature Phase Shift Keying
R4	Release 4 (3GPP)
R5	Release 5 (3GPP)
R99	Release 1999 (3GPP)
RACF	Radio Access Control Function
RACH	Random Access Channel
RACS	Resource and Admission Control Subsystem
RADIUS	Remote Authentication Dial In User Service
RAN	Radio Access Network
RANAP	Radio Access Network Application Part
RAS	Radio Access Station
RCH	Random Access Channel
RF	Radio Frequency
RFID	Radio Frequency Identification
RFC	Request for Comments
RLC	Radio Link Control
RLS	Remote List Server
RNC	Radio Network Controller
RRM	Radio Resource Management
RSA	Rivest, Shamir and Adleman
RSN	Robust Security Network
RSVP	Resource Reservation Protocol
RTCP	Real-time Transport Control Protocol
RTP	Real-time Transport Protocol and Real Time Polling (WiMAX)
RTS	Request To Send
RTSP	Real Time Streaming Protocol
SACCH	Slow Associated Control Channel
SCCH	Single Cell Control Channel
S- CSCF	Serving Call Server Control Function
SDH	Synchronous Digital Hierarchy
SDK	Software Developer's Kit
SDP	Session Description Protocol
SDU	Service Data Unit
SG	Signalling Gateway
SGSN	Serving GPRS Support Node
SHA	Secure Hash Algorithm
S/I	Signal to Interference ratio
SIFS	Short InterFrame Space
SIM	Subscriber Identity Module

SIP	Session Initiation Protocol
SM	Session Management
SME	Short Message Entity
SMG	Special Mobile Group
SMS	Short Message Service
SMTP	Simple Mail Transfer Protocol
SNR	Signal to Noise Ratio
SPDF	Service Policy Decision Function
SS7	Signalling System Number 7
SSID	Service Set Identifier
SSL	Secure Socket Layer
SSP	Service Switching Point
SWG	Signalling Gateway
TCP	Transmission Control Protocol
TCP/IP	Transmission Control Protocol/Internet Protocol
TDD	Time Division Duplex
TDM	Time Division Multiplex
TDMA	Time Division Multiple Access
TE	Terminal Equipment (ETSI Committee)
TLS	Transport Layer Security
TMSI	Temporary Mobile Subscriber Identity
TOS	Type of Service
TPC	Transmitter Power Control
TR	Technical Report (ETSI)
TS	Technical Specification (ETSI)
TSPec	Traffic Specification
TTI	Transmission Time Interval
TXOP	Transmission Opportunity
UDP	User Datagram Protocol
UGS	Unsolicited Grant
UHF	Ultra High Frequency
UICC	UMTS Integrated Circuit Card
UMA	Unlicensed Mobile Access
UMTS	Universal Mobile Telecommunications System
UNC	UMA Network Controller
UNI	User-to-Network Interface
URL	Uniform Resource Locators
USB	Universal Serial Bus
USIM	User Services Identity Module
UTRAN	UMTS Terrestrial Radio Access Network
UWB	Ultra Wide Band
VC	Virtual Circuit
VCC	Voice Call Continuity

VHE	Virtual Home Environment
VLR	Visitor Location Register
VoIP	Voice over IP
VPN	Virtual Private Network
WAN	Wide Area Network
WAP	Wireless Application Protocol
WCDMA	Wideband Code Division Multiple Access
WECA	Wireless Ethernet Compatibility Alliance
WEP	Wired Equivalent Privacy
WiFi	Wireless Fidelity
WiMAX	Worldwide interoperability for Microwave Access
WLAN	Wireless Local Area Network
WLL	Wireless Local Loop
WMM	Wireless MultiMedia
WPA	WiFi Protected Access
WRC	World Radiocommunication Conference
WTP	Wireless Termination Points
WWW	World Wide Web
XML	eXtensible Mark-up Language
XMPP	Extensible Messaging and Presence Protocol

Chapter 1: Introduction

Not another book on mobile? Come on Wisely, how an earth can you justify yet another book with 4G in the title? I suppose the best answer is that I have never seen a book quite like this one! In truth it grew out of a short, two day, course on the future of mobile that myself and a few colleagues have given at Oxford for the last few years. The course was intended as an overview of the key up and coming technologies as well as commentaries from leading experts in each field as to why they were important and what the key issues were. In trying to find a book to support the course it proved impossible to find something that was both comprehensive – covering all the topics – but also having sufficient detail to avoid missing the key facts about the technologies. There are plenty of books dedicated to (say) the IMS or 3G but they are jam-packed with the nitty gritty detail of the protocols – required reading if you are working in the area, but if you just want a quick overview then they are very hard work indeed. Overview books seemed equally unsatisfactory; suffering from being very unevenly written – with each chapter often written by different authors of varying quality and consistency. Worse still was the "Spangliase effect" – an affectionate term by one of my colleagues for English prose written by native Spanish speakers that seems to be grammatically correct but just sounds odd and is hard to read fluently. In addition you really got only a very superficial level of detail – which often didn't offer much insight into the technology and its shortcomings (such as why Wireless LANs (WLANs) have such high power consumption and why it is so hard to do anything about it). Both types of books were also weak on the commercials – lots about the technology but no commentary as to whether it would be a success or what the leading operators really thought about it. For that you really need to attend some of the expensive non-academic conferences that are run by the likes of IIR or Marcus Evans. If you are not a speaker then they can cost upwards of £2000 a go. You also probably need an insight from key operators and vendors – things they are very wary of talking about in public.

IP for 4G   David Wisely
© 2009 John Wiley & Sons, Ltd.

So, to summarise the case for the defence, this book attempts to cover all the key technologies in the mobile space at a decent level of detail – between the horror of describing all the protocols and a shallow summary. I have also tried to tackle the key commercial questions and offer an opinion on the main issues in mobile today. Lots of people should read this book – I need the money! – no this book is really for people in the industry who want an update into what's happening, for people with an engineering background who want to know how mobile comms work at a deeper level and for all those with an interest in mobile comms over the next few years.

So what of your author? Well I have been in telecoms for 20 years at BT. I have worked in mobile for 10 years – looking at WLANs before they were popular, and working, for the last few years, on convergence between fixed and mobile. I was technical manager for the EU BRAIN/MIND projects and am now heading up BT's convergence research. In relation to bringing you this book I have been to many conferences – boring audiences from Moscow to Melbourne – read many dull books and scoured the Internet for news and useful links. I have endured a dodgy shoulder (leaning over tables), vibration white finger (bad typing habits) and pushy salesmen from mobile vendors. In order to survive all this you need a very good sense of humour and you will find I have included a few jokes along the way to lighten things up.

If you have an engineering background then you can easily follow the book – you might need a primer on IP or Mobile radio and GSM – but there are a couple listed at the end of the chapter – although most of the points on mobile are included in the book. If you don't have an engineering background then there is still plenty of commercial insight to be had and I have tried to isolate the more detailed sections and make them as self-contained as possible.

The book follows on a bit from *IP for 3G* (Wisely *et al.*) which was published in 2002 – which covered IP more thoroughly with chapters on IP Mobility, IP QoS and SIP (Session Initiation Protocol). Things have moved on a lot in mobile – in my view this is a very exciting time for mobile as this chapter explains – with IP QoS and mobility proving to be rather "red herrings" – in that mobile standards haven't really used them and the IETF[1] has been very slow at moving these on. It is rather the air interfaces – WLAN, WiMAX, HSPA, LTE[2] – that are the buzzwords for the industry and that is how I have chosen to structure the chapters. They sort of go in an order of 3G, WLANs, 3.5G, WiMAX and then a couple of chapters on service creation (IMS) and mobile services themselves. I apologise for readers who already own *IP for 3G* because the chapter on 3G is lifted straight from that book – in order to make this book complete in its own right (if you feel hard done by then email me and I will send you a copy of my Oxford slides that accompany the course for free).

---

[1] Internet Engineering Task Force – the experts who make up the Internet protocols.

[2] WLAN – Wireless LANs, HSPA (high speed packet access – next phase of 3G – sort of 3.5G), WiMAX – Worldwide Interoperability for Microwave Access (sort of wide area WLAN) and LTE – long term evolution, the mobile industry equivalent of WiMAX. Much more about these later.

The rest of the introduction is really setting the scene – it includes a prologue that looks at the history of mobile (very briefly – thank goodness I hear you say), introduces the props (the key technologies) as well as the main characters. After that it is up to you in what order you read the book, but it has been written to be read linearly and the later chapters assume you know some of what precedes them. I hope you enjoy reading the book and that at least something in it is useful or makes you smile.

1.1 Prologue – The Generation Game

Cellular mobile has been around for over 30 years – Figure 1.1 shows a 1970s scene with what might be called a zero generation mobile. Things have moved on a lot since then – not only in fashion and hairdressing – with the launch of analogue cellular systems in 1980 (1G). Such systems were characterised by their lack of roaming – many countries developed their own system – lack of security and poor battery life. Second generation (2G) digital systems were launched in 1990. Several systems (see Table 1.1) were launched but it has been GSM, 18 years later, that has come to dominate mobile communications around the world. As of March 2008 there were a little over 3 billion GSM phones in use around the world – representing about 85% of the total. 3G, by contrast had only 350 million or so subscribers (although rising fast) (Ref. 1). It is also reckoned that GSM will not be switched off until 2018 at the earliest in the UK alone. It could almost be argued that the whole success of mobile is based on GSM – standardisation, volumes and falling costs driving a virtuous circle of success.

Figure 1-1. Zero generation mobile. (*Source:* BT Heritage. Reproduced by permission of BT Heritage © British Telecommunications plc.)

Systems	Products
Cellular 1G — 1980 launch	— advanced mobile phone system (AMPS — USA) — total access communication system (TACS — Europe) — analogue voice, insecure, no roaming
Cellular 2G — launched in Europe 1990	— IS-136 (TDMA) and IS95 (CDMA) — USA — GSM (TDMA) — Europe — personal digital cellular (PDC) — Japan — digital voice, roaming, low-rate data
Cellular 2.5G — 1996	— general packet radio services (GPRS) — GSM enhancement — 20-64 kbit/s — enhanced data rates for global GSM evolution (EDGE — 100 kbit/s — higher rate data)
3G — 2001 — BT launch UMTS on Isle of Man (3G)	— mobile Internet — 100 kbit/s

Table 1-1. Mobile generations.

2G was about voice – good quality mobile voice with roaming, secure encryption and improved talk and stand-by times. SMS – the now ubiquitous messaging service – happened almost by accident, since it was originally intended as an internal communication system for GSM engineers. Even today voice and SMS accounts for over 90% of the typical non voice revenue of UK mobile operators (Ref. 2). General packet radio system (GPRS) – often called 2.5G – was introduced to allow (nearly) always-on data connection and volume-based charging for data sessions. Previously, users of wireless application protocol (WAP) – a menu-based, cut-down, version of HTML specially designed for handsets – had to have a circuit connection (with time-based charging) open for the duration of the session. GPRS offers data users up to about 50 kbit/s – something like dial-up rates – which improves the WAP experience. 3G was launched in Europe on the Isle of Man by BT Cellnet in 2001 and was sold as offering users communication "any time, any place, anywhere" – the so called Martini effect. It was intended to offer high-speed data services, video telephony, lower costs and a host of novel services as well as the "mobile Internet". In Europe at least, however, take-up has been slower than predicted with currently 8% of UK mobiles being 3G (some 5 million out of 65 million [Ref. 3]) with most used only for voice and SMS. Only now are we seeing a move to mass market mobile data as enhancements to 3G's data capability have been rolled out and prices dramatically reduced.

1.2 The Props – WLAN, WiMAX and All That

Figure 1.2 shows (admittedly not the best version) of the most famous diagram in mobile. However, it is a very useful starting point for understanding why

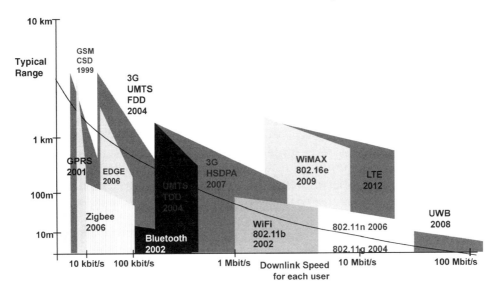

Figure 1-2. Mobile technologies – the most famous diagram in mobile (BT version). (Source: BT. Reproduced by permission of © British Telecommunications plc.)

there are so many mobile technologies. The fundamental thing about radio is that there is always a trade-off between range, bitrate and mobility. Imagine you have been given a chunk of spectrum (maybe you bought it in a government auction) and you want to get busy in the shed and build a small mobile radio system that you and your community can use. There will always be some power constraint on how much power you can transmit – either because of restrictions on the spectrum (someone else shares it) or the battery power on the mobile terminals or just the cost of the power amplifiers on the base-stations. You then need to decide what bandwidth the users need – this will be much higher for video than for voice say. Let's, for the sake of argument, say you want video – 1Mbit/s for good quality on a PDA maybe. This will set the range of your system – in conjunction with maximum transmit power and the frequency of the spectrum (the lower the frequency the better the range). If you had 10MHz of spectrum you could give your 10 neighbours 1MHz of dedicated spectrum. But then more people want to join and you soon realise that most of the time the spectrum is unused – so you decide to have a pool of spectrum resources and allocate it as users actually want to use it – 100 people can join as, statistically, only a few of them use it at any one time. Then more people want to use it – a bit of research shows that fancy codecs and compression of the video means that users don't really need a 1Mbit/s video stream. They can send only the parts of the scene that change – suitably compressed. This has a bursty traffic profile – when the scene is slowing changing there is little traffic – so you can multiplex many more users – say 1000. But now Quality of Service is an issue – you need complicated protocols to signal, admit and police requests because the system capacity is near to maximum sometimes and the bandwidth required fluc-tuates. Then more people want to join – so you need to upgrade the system but this is going to cost and so you want to start charging. However, some people pay but others are freeloading – so you need a security system for authenticating and authorising users. Then, with all this cash, you want to extend the system to people on their bikes and in their cars. You soon notice that your coverage range is much reduced as the mobility of the users rises (basically because of Doppler and other radio effects associated with movement). In order to keep the coverage area – after much debate with the vicar about mammon and greed in society – you pay him a hefty whack to put up another base-station on his church and pay BT to connect it to your servers. However, users moving from the coverage area of one base-station to another want seamless operation – so you have to engineer a handover solution and make sure the cells overlap but not on the same frequency otherwise they interfere. Congratulations – you are now a mobile operator.

What this fanciful romp shows is that mobile systems have a number of attributes that are trade-offs:

- Data rate;
- Capacity;
- Coverage;
- Mobility;
- Frequency planning.

And some that are optional in that they are not essential but desirable:

- Quality of service support;
- Mobility support;
- Security;
- Battery life;
- Handover support.

Overall the laws of physics don't change – which is a pity as there are quite a few "interesting things" I'd like to do that contravene them. What really happens, as technology evolves, is that Moore's law and electronics/software advances and gets cheaper to the point where mobile systems are better able to approach the fundamental limits of capacity that are set by the laws of physics.[3]

---

[3] Such as Shannon's law which basically says the capacity of a comms channel is proportional to the bandwidth of the channel and decreases as the noise increases. In mobile systems this noise mostly comes from other users – within the same or neighbouring cells. Modern systems – like WiMAX and LTE are said to approach the Shannon limit.

You can see this in Figure 1.2 where I have drawn a hyperbola to show that for most systems bandwidth*range = constant. The constant goes up in time – as electronics and software advances and becomes cheap enough for mobile systems – but slowly. I reckon the equivalent Moore's law constant for mobile (i.e. how long it takes mobile efficiency to double, all other things being equal) is about five years – i.e. the capacity of 1MHz of spectrum doubles every five years and quadruples every 10 years or one mobile generation.

So the different technologies you will read about in the next four chapters (3G, HSPA, LTE, WLAN and WiMAX) offer different solutions to the above functions and different trade-offs. Apart from WLANs they offer complete solutions – with a fully integrated architecture and tightly coupled functions. WLANs are the odd man out in that they provide only a minimal set of functions but very high data rates over a short range. The WLAN chapter will explore the consequences of this and explain how this came about.

1.3 The Players – The Jester, the Harlequin and the Guy with the Beard and Sandals

The players – well they are a mixed assortment. Firstly, in order of importance I suppose, come the mobile operators (MNOs they are often called – Mobile Network Operators). The MNOs are the control freaks of the industry. They sell most of the phones – often in their own shops. They control authentication through the SIM card. They also are the guardians of the services that you are or are not allowed to have on your handset and/or use on their network. This comes about because most handsets are subsidised (in Western markets particularly – even Apple is now allowing the iPhone to be subsidised). When you get one from a MNO the deal is that the handset is cheap but it is loaded with applications specific to that operator and the whole phone is then "locked down" to prevent alteration of the settings. Worse still the handset is locked to the network so that you can't just buy a SIM from another operator and drop it in your existing handset. Mobile operators hate that sort of thing. They also have battled long and hard against the Internet business model (Figure 1.3) – launching their own

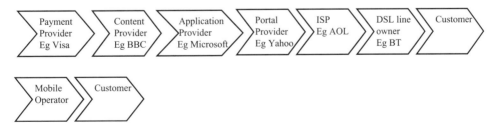

Figure 1-3. Comparing the Internet (top) and Mobile (bottom) value chains. (Source: Author. Reproduced by permission of © Dave Wisely, British Telecommunications plc.)

"Mobile Internets" – which were in fact walled gardens of sites that the MNO has selected (i.e. mostly those that generated a rake-off). The MNOs are still in land grab phase in the developing world – signing up armies of new customers for voice. In the developed world the game is to offset static user numbers and falling voice revenue with new data services.

You have to feel sorry for the mobile equipment manufacturers. It has been said that since the .com bust the total number of Telecoms jobs has shrunk by 500 000. It is the mobile vendors who have taken the biggest hit with major mergers – such as Alcatel-Lucent and Nokia-Seimens – showing the consolidation going on in the industry. Mobile vendors are under threat from newer rivals from China and the Far East – such as Huawei and many new handset vendors – who are pushing down prices. Vendors are very keen that MNOs upgrade their networks with the next version that has been carefully worked on by both of them in standards. They are not keen on "left field" initiatives from the computer industry vendors to offer different systems in the form of WiMAX and WLANs. The computer industry – led by companies such as Intel – however, has been busy adding functionality to WLANs and developing WiMAX – which has been called "WLAN on steroids". If all the laptops and PDAs have WLAN and WiMAX in them then that is a huge boost for these technologies over the more traditional systems from mainstream mobile vendors.

Spare a thought for the customers who, ultimately, have to pay for all this. Customers really like mobile phones and they really like chatting on them. If you live in a developed country and haven't got at least one mobile then most people assume you are either a hippie or living on the streets. People have been paying a "mobile premium" of 10–100 times for voice (over and above what they would have paid on a fixed phone) for the privilege of being mobile. Even today this is still 2–10× for voice (in-country) and much higher when roaming. Most users say they would talk for longer if tariffs were cheaper – which they have been getting for the past few years. Users have also taken to texting (luv it or h8 it) which has also been a massive hit. Users have not taken to video-telephony or picture messaging and are just showing an interest in the Mobile Internet.

Then we have the standards guys (I say guys because 95% of them are men). Standards are very dull – very long tedious meetings in dreary locations arguing about minutiae and eating too much. However, without standards there is no way the mobile phone revolution could possibly have happened. It has lowered prices by an order of magnitude compared to the fragmentary diverse standards of first generation that were different for pretty much every country. GSM now has 85% of the world market with handset prices falling year on year as volumes have ramped up. There are two standards "camps" that we will be mainly concerned with. The mobile camp is dominated by 3GPP and 3GPP2 with various industry for an ancillary to this. Then there is the computer camp with the IEEE responsible for WiMAX and WLAN standards. The IETF (Internet Engineering Task Force) – otherwise known for their resemblance to 1970s heavy metal bands (all beards and sandals) – are responsible for IP standards.

Finally the industry police (regulators) have a major role in the play. They control the spectrum – the "fuel of mobile". They also have the power to set prices and to order additional competition. This is not the place for a debate about the merits of regulation but its effect can be seen in the price of DSL across Europe where, as we shall see, it has a major effect on the possible future of mobile.

A mention of the bit-players perhaps? The fixed operators are busy trying to deliver something called convergence – where the fixed and mobile worlds become part of the same telecom experience. They are mostly just trying to avoid fixed to mobile substitution which is gaining traction – in Austria 20% of people don't have a fixed line. Then there are the Internet giants – Google, Amazon, Microsoft, Wisely.com – who are looking at the mobile space as a way to break out of the fixed Internet. And what about the researchers? – underpaid, locked away in dingy labs, no appreciation . . . .

1.4 The Plot – The Elephant in the Room

The elephant in the room is the Internet. The Internet is now everything in communications – despite the fact that there are four mobile phones for every PC (a fact the mobile vendors are very keen to tell you) – this does not reflect where the economic power lives. Ninety-five per cent of those handsets are used for voice and text and less than 1% have been used to book tickets or watch videos. The power of the Internet giants (Amazon, Google . . .) is so large, as we noted above, that the mobile operators have started to give up trying to build a separate mobile Internet and are partnering with them.

IP is the key to the Internet and to understanding where mobile is going – in my view anyway. Hence the title IP44G even though developments in IP itself are only a minor part of the book (in fact IP hasn't changed much in the last five years – IPv6 was just over the horizon then and is just over the horizon now). It is worth remembering what IP really is and why it has been so successful because this is what is driving the likes of LTE and WiMAX as well as the IMS (Internet Multimedia Subsystem – IP service creation for mobile networks as explained in Chapter 6). Firstly, let's think what IP is – Figure 1.4 shows the standard IP stack – which hopefully brings some recognition? The key point about IP is that you can put it over anything – any layer 2 technology from ATM to Ethernet to carrier pigeons (there was an Internet draft about IP over carrier pigeons (Ref. 4)). The other key point about IP was that you could put anything in the IP packets – music, voice, gaming data – and the packets were treated just the same by all the routers. This is called the transparency/layering principle and is very powerful and meant that all the services and control (old and new) was de-coupled from basic IP operation. "IP over everything and everything over IP". IP is the bottleneck and the stack might be more appropriately drawn as in Figure 1.5.

The other big design principle of IP is the "end-to-end" principle – meaning that end points are the best place to put functionality whilst the network should be dumb – just transporting IP packets around but never opening them or performing functions on them other than routing them to their final destination.

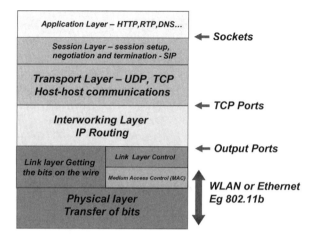

Figure 1-4. The IP stack (Source: Author. Reproduced by permission of © Dave Wisely, British Telecommunications plc.)

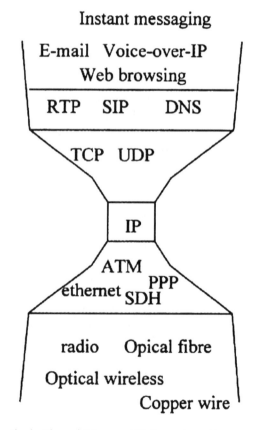

Figure 1-5. IP as the bottleneck (Source: BT. Reproduced by permission of © British Telecommunications plc.)

It is important because it allows IP systems to be very flexible, to do many different tasks simultaneously (from serving web pages to voice flows to downloading files and so on) and to do new things without changes being required in the network. The IP architecture is very loose – with one protocol per function and no coupling between them.

2G mobile has none of these characteristics! It is what is known as a stovepipe solution – meaning every part of the stack is interdependent on every other part. This made sense when the service set was very limited – in the case of GSM it was just voice (SMS was originally for engineers to signal to each other). Each part of the GSM stack was optimised for voice and the result was far more efficient than a voice over IP system could ever have achieved. But, and this is the big but, when it came to IP services GSM was next to useless. It was also very hard to upgrade or change even for voice – offering higher voice quality (so that you could listen to CD quality music say) was impossible because that would have necessitated changing all the base-stations, handsets and so on. Even trivial new services – such as short code dialling (e.g. 901 to get my messages) took a long time to appear and needed a new standard and major upgrades to the network just to make it work for roaming customers. Figure 1.6 shows the general idea.

3G was born before the Internet. As we shall see in the next chapter 3G designers opted for a revolutionary air interface and an evolutionary network. The result: 3G is very good for voice but very bad for IP. It is bad for IP firstly because the air interface doesn't cope well with IP applications – the throughput and capacity is limited in relation to current IP applications. The latency (end to end delay) is also high (300–500 ms) compared to DSL as is the time taken to start a browsing session from idle of a couple of seconds. Secondly, there is nothing in the standards that allows the MNOs to create and control Internet services – they are reduced to a bit carrier with the "value-add" escaping to the likes of Google – whilst this is perfectly in line with the Internet business model it is not well aligned with the existing mobile business model – that covers services and content as well as connectivity.

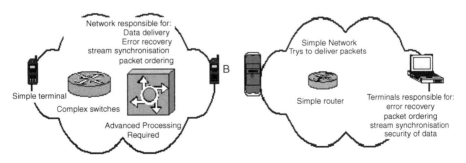

Figure 1-6. The IP way and the Telco way. (Source: BT. Reproduced by permission of © British Telecommunications plc.)

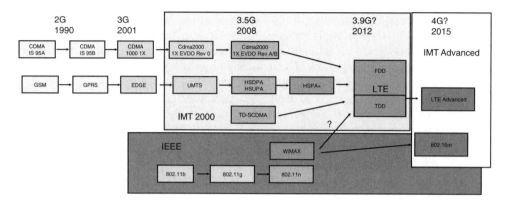

Figure 1-7. The road to 4G. (*Source:* Author. Reproduced by permission of © British Telecommunications plc.)

So 3G was never going to be the solution for the mobile Internet. Two solutions to carrying IP packets at DSL-like data rates came forward. The computer industry came up with WLANs – for connection to portable computers – PDAs and laptops. The mobile industry came up with 3.5G – enhancing the existing infrastructure and the 3G air interface in particular to better cope with IP. The two camps have also developed WiMAX – once called "WLAN on steroids" – and LTE (3.9G) – see Figure 1.7. Both are very similar in having a very efficient air interface that is optimised for IP and takes advantage of electronic progress. Both technologies feature "all-IP" networks that are "flatter" (i.e. have fewer layers) and more like routed IP networks.

The mobile industry also came up with the IMS – a platform for service creation and control that would allow them to create services and capture value from Internet services (such as Voice over IP, video calling and Instant Messaging).

So, you see, everything in mobile is being driven by the need to carry IP and IP services. The Internet is now so big and so valuable that if the mobile industry is to move on from voice and messaging then it will have to be through IP. It doesn't matter that the MNOs may or may not save money with IP or that it is or is not technically better – it is simply the power and size of the Internet that is driving mobile forward. People will pay a mobile premium for a service they use in the fixed world – they paid a huge premium for mobile voice. If mobile operators follow the same paradigm of taking a successful service and making it mobile then they may expect at least some premium for the mobile Internet. Whether they capture any further value is a moot point and one of my top 10 questions posed below.

1.5 Epilogue – The Road to 4G

I think this is a very exciting time to be involved in the mobile industry. With HSPA finally offering some of the promise of 3G as a technology capable of delivering mobile data and Wireless Cities – whole centres covered by WLANs. Then there are the new radio systems – WiMAX and LTE – with "magic bullet" technologies like OFDM (Orthogonal Frequency Division Multiplexing) and smart antennas – systems that promise to be all-IP from the start.

With so much hype and conflicting messages from different parts of the industry I would like to set 10 key questions that we can return to in the conclusion to see if the book has shed any light on:

1. Why has 3G been so slow to take-off?

2. What really is the Mobile Internet?

3. How good is 3.5G?

4. Why are WLANs so limited (range, power consumption etc.) and is that going to change?

5. Is all the hype around WiMAX justified?

6. Is LTE just the same as WiMAX?

7. Will the fixed Internet business model be carried over into the mobile space?

8. How will mobile operators create attractive services of their own? And will they succeed?

9. What services (if any) – over and above voice and messaging and the Internet– do users really want?

10. How much are users willing to pay for new services?

All of these questions are linked to the issue of "what is 4G?" In some sense that is the main theme of the book as we follow the attempts of the mobile and computer worlds to deliver mobile Internet. Of course it is quite hard to define a mobile generation (as there have been only three) but what has distinguished them so far has been:

- A gap of about 10 years;

- A completely new air interface and network;

- New terminals;

- About 4 times the capacity of the previous generation;

- New spectrum.

On this metric only LTE and, possibly WiMAX qualify as new generations. But not everybody buys this argument (Ref. 5). Some people believe that converged solution – with many different access technologies such as: WLAN, WiMAX, 2G and 3G is what 4G is really about. Others say it is defined by a service set. All the chapters in the book build towards an answer – which is revealed in the final chapter after the above questions have been revisited.

The ITU (International Telecommunications Union – see the who's who of standards in the next chapter) have defined 4G as 100Mbit/s + – and on that scheme neither LTE nor WiMAX qualify. Indeed LTE has been officially registered with the ITU as a 3G technology (insiders call it 3.9G) with something called "LTE-advanced" destined as the 4G version. Much of this can be put down to the industry trying not to let 3G seem dated before it has achieved a significant market share (around the world 3G has less than 10% of all connections) and an attempt to talk down spectrum prices in auctions currently taking place. 4G is still a confusing term in the industry and one of the topics we will return to when we have seen what some of these technologies have to offer.

If you don't know about the history of 3G and how it works then I suggest you shuffle off to Chapter 2 and find out. If you know all about mobile then move directly to Chapter 3 on WLANs – if you know all about those then you are a smart Alec who should be on Mastermind and not reading this book!

References

[1]   GSM World — http://www.gsmworld.com/technology/gsm.shtml
[2]   Informa Telecoms and Media: "Mobile industry outlook", (2006) – http://shop.informatm.com/
[3]   Informa Telecoms and Media (including EMC Database, Global Mobile) – http://shop.informatm.com/
[4]   IP over Avian Carriers (IPoAC) – RFC 1149. On 28 April 2001, IPoAC was implemented by the Bergen Linux user group who sent nine packets over a distance of three miles each carried by a pigeon – they received four responses.
[5]   Pereira, J. "Fourth Generation: Now, it is Personal", Proceedings of 11th International Symposium on Personal, Indoor and Mobile Radio Communication, IEEE, 2, pp. 1009–1016 (September 2000).

More to Explore

Radio Tutorial
"Wireless communications — the Fundamentals", Terry Hodgkinson *BT Technology Journal* ISSN 1358-3948 (Print) 1573-1995 (Online). Issue Volume 25, Number 2/April, 2007.

Overview of GSM
– try http://www.iec.org/online/tutorials/gsm/
Or http://www.palowireless.com/gsm/tutorials.asp which has a handy list of web resources on GSM tutorials.

IP design
– RFC 1958 – "Architecture Principles of the Internet", Carpenter, B. *et al.*, June 1996 and "End-to-End Arguments in System Design".

Chapter 2: An Introduction to 3G Networks

2.1 Introduction

So what exactly are 3G networks? 3G systems might be defined by: the type of air interface, the spectrum used, the bandwidths that the user sees or the services offered. All have been used as 3G definitions at some point in time. Currently there are only three flavours of 3G – known as UMTS (Universal Mobile Telecommunications System – developed and promoted by Europe and Japan), cdma2000 (developed and promoted by North America) and TD-SCDMA (Time Division-Synchronous Code Division Multiple Access) developed by the Chinese. All are tightly integrated systems that specify the entire system – from air interface to the services offered. Although each has a different air interface and network design they will offer users broadly the same services of voice, video and fast Internet access.

3G (and indeed existing second generation systems such as GSM) systems can, very crudely, be divided into three (network) parts: the air interface, the radio access network and the core network. The air interface is the technology of the radio hop from the terminal to the base station. The core network links the switches/routers together and extends to a gateway linking to the wider Internet or public fixed telephone network. The Radio Access Network (RAN) is the "glue" that links the core network to the base-stations and deals with most of the consequences of the terminal's mobility.

The purpose of this chapter is really to highlight the way UMTS (as an example of a 3G system) works at a network and air interface level – in terms of mobility management, call control, security and so forth. This sets the scene to

IP for 4G   David Wisely
© 2009 John Wiley & Sons, Ltd.

see how WLANs do things differently (Chapter 3), how 3G systems are being upgraded (Chapter 4) and how new systems like WiMAX (Chapter 5) and LTE (Chapter 4) are being engineered to be all-IP systems.

The history of 3G development shows that the concepts of 3G evolved significantly as the responsibility for its development moved from research to standardisation – shedding light on why 3G systems are designed the way they are. Included in this section is also a "who's who" of the standards world – a very large number of groups, agencies and fora have, and still are, involved in the mobile industry. In the second half of the chapter we introduce the architecture of UMTS (the European/Japanese 3G system) and look at how the main functional components – QoS, mobility management, security, transport and network management – are provided. A short section on the US cdma2000 3G system is also included at the end of the chapter.

2.2 Mobile Standards

2.2.1 Who's Who in Mobile Standards

Before we start delving into the labyrinthine world of standards it is, perhaps, a good point to provide a brief "who's who" to explain recent developments in the standards arena.

- 3GPP – www.3gpp.org

  In December 1998 a group of five standards development organisations agreed to create the Third Generation Partnership Project, these partners were: ETSI (EU);ANSI-TI (US); ARIB and TTC (Japan); TTA (Korea) and CWTS (China). Basically this was the group of organisations backing UMTS and, since August 2000 – when ETSI SMG (Special Group Mobile) was dissolved – has been responsible for all standards work on UMTS. 3GPP were responsible for completed the standardisation of the first release of the UMTS standards – Release 99 or R3. GSM upgrades have always been known by the year of standardisation and UMTS began to follow that trend – until the Release 2000 got so bogged down and behind schedule that it was broken into two parts and renamed R4 and R5. As of Q2 2008 3GPP has completed R7 but not yet finalised R8. In Chapters 4 and 6 we will see the incremental changes that these new releases have brought to the network, air interface and terminals. In this chapter we only describe the completed R3 (formally known as Release 99) as this is what has been deployed and has only recently, in the last year or so, begun to be updated.

- 3GPP2 – http://www.3gpp2.org/

  3GPP2 (www.3gpp2.org) is the cdma2000 equivalent of 3GPP – with ARIB and TTC (Japan), TR.45 (US) and TTA (Korea). It was responsible for standardising cdma2000 based on evolution from the cdmaOne system and using an evolved US D-AMPS network core. (The latter part of this chapter gives an account of packet transfer in cdma2000).

- ITU – http://www.itu.int/net/home/index.aspx

  The International Telecommunications Union (ITU – www.itu.int) was the originating force behind 3G with the FLMTS concept (pronounced Flumps and short for Future Land Mobile Telecommunication System) and work towards spectrum allocations for 3G at the World Radio Conferences. The ITU also attempted to harmonise the 3GPP and 3GPP2 concepts and this work has resulted in these being much more closely aligned at the air interface level. Currently the ITU is developing the concepts and spectrum requirements of what it calls 4G and sits beyond LTE and WiMAX.

- IETF – http://www.ietf.org/

  The Internet Engineering Task Force (www.ietf.org) is a rather different type of standards organisation. The IETF does not specify whole architectural systems, rather individual protocols to be used as part of communications systems. IETF protocols such as SIP (Session Initiation Protocol) and header compression protocols have been incorporated into the 3GPP standards. IETF meetings happen three times a year and are: completely open, very large (1000 + delegates) and less process driven (compared to the ITU meeting say). Anyone can submit an Internet draft to one of the working groups and this is then open to comments – if it is adopted then it becomes a Request For Comments (RFC), if not then it is not considered further.

- ETSI – http://www.etsi.org/WebSite/homepage.aspx

  ETSI (the European Telecommunications Standards Institute) is a nonprofit making organisation for telecommunications standards development. Membership is open and currently stands at 789 members from 52 countries inside and outside Europe. ETSI is responsible also for DECT although GSM developments have been passed to 3GPP.

- FMCA – http://www.thefmca.com/

  The Fixed-Mobile Convergence alliance. This is an organisation of operators (fixed and mobile) and leading vendors who are working on product specifications for converged, fixed-mobile, devices. We will talk a lot more about convergence in Chapter 6.

- OMA – http://www.openmobilealliance.org/

  Open Mobile Alliance. This consists of vendors, mobile operators, content providers and IT companies and aims to "grow the market for the entire mobile industry by removing barriers to interoperability, supporting a seamless and easy to use mobile experience for users and a market environment that encourages competition through innovation and differentiation".

- IEEE – http://www.ieee.org/portal/site

Institute of Electrical and Electronics Engineers (although now called the IEEE). Its website proclaims that it is "A non-profit organization, IEEE is the world's leading professional association for the advancement of technology". The IEEE is responsible for WLAN and WiMAX standards and represents the computer industry approach to mobile.

- WiMAX Forum – http://www.wimaxforum.org/home/

The WiMAX forum. "is an industry-led, not-for-profit organization formed to certify and promote the compatibility and interoperability of broadband wireless products based upon the harmonized IEEE 802.16/ETSI HiperMAN standard". It is responsible for WiMAX certification and interoperability as well as all nonradio network aspects.

2.3 The History of 3G

It is not widely known that 3G was conceived in 1986 by the ITU (International Telephony Union). It is quite illuminating to trace the development of the ideas and concepts relating to 3G from conception to birth. What is particularly interesting, perhaps, is how the ideas have changed as they have passed through different industry and standardisation bodies. 3G was originally conceived as being a single worldwide standard and was originally called FLMTS (pronounced Flumps and short for Future Land Mobile Telecommunication System) by the ITU. By the time it was born it was quins – five standards – and the whole project termed the IMT-2000 family of standards. After the ITU phase ended in about 1998 two bodies – 3GPP and 3GPP2 – have completed the standardisation of the two flavours of 3G and are actually being deployed today and over the next few years (UMTS and cdma2000 respectively).

It is convenient to divide up the 3G gestation into three stages (so-called trimesters):

Pre 1996 – The Research Trimester

1996-1998 – The IMT-2000 Trimester

Post 1998 – The Standardisation Trimester

Readers interested in more details about the gestation of 3G should see Ref. 1– which describes who introduced the three stages).

2.3.1 Pre-1996 – The Research Trimester

Probably the best description of the original concept of 3G is to quote Alan Clapton – head of BT's 3G development at the time "3G . . . The evolution of mobile communications towards the goal of universal personal communications, a range of services that can be anticipated being introduced early in the next century to provide customers with wireless access to the information super highway and meeting the 'Martini' vision of communications with anyone, anywhere and in any medium" (Ref. 2).

Here are the major elements that were required to enable that vision:

- A worldwide standard

    At that time the European initiative was intended to be merged with US and Japanese contributions to produce a single worldwide system – known by the ITU as FLMTS. The vision was a single handset capable of roaming from Europe to America to Japan.

- A complete replacement for all existing mobile systems

    UMTS was intended to replace all second generation standards, integrate cordless technologies as well as satellite (see below) and also to provide convergence with fixed networks.

- Personal mobility

    Not only was 3G designed to replace existing mobile systems, its ambition stretched to incorporating fixed networks as well. Back in 1996, of course, fixed networks meant voice and it was predicted in a European Green Paper on Mobile Communications (Ref. 3) that mobile would quickly eclipse fixed lines for voice communication. People talked of Fixed Mobile Convergence (FMC) with 3G providing; a single bill, a single number, common operating and call control procedures. Closely related to this was the concept of the Virtual Home Environment (VHE).

- Virtual Home Environment

    The virtual home environment was where users of 3G would store their preferences and data. When a user connected, be it by mobile or fixed or satellite terminal, they were connected to their VHE which then was able to tailor the service to the connection and terminal being used. Before a user was contacted then the VHE was interrogated – so that the most appropriate terminal could be used and the communication tailored to the terminals and connections of the parties.

- Broadband service (2Mbit/s) with on-demand bandwidth

    Back in the early 1990s it was envisaged that 3G would also need to offer broadband services – typically meaning video and video telephony. This broadband requirement meant that 3G would require a new air interface and this was always described as broadband and typically thought to be 2Mbit/s. Associated with this air interface was the concept of bandwidth on demand – meaning it could be changed during a call. Bandwidth on demand being used, say, to download a file during a voice conversation or upgrade to a higher quality speech channel mid-way through a call.

- A network based on B-ISDN (Broadband ISDN)

Back in the early 1990s another concept – certainly at BT – was that every home and business would be connected directly to a fibre optic network. ATM transport and B-ISDN control would then be used to deliver broadcast and video services: an example being video on demand whereby customers would select a movie and it would be transmitted directly to their home. B-ISDN was supposed to be the signalling for a new broadband ISDN service based on ATM transport – it was never actually developed and ATM signalling is not sufficiently advanced to switch circuits in real time. ATM – asynchronous transfer mode – is explained in the latter part of this chapter: it is used in the UMTS radio access and core networks. Not surprisingly, given the last point it was assumed that the 3G network would be based on ATM/B-ISDN.

- A satellite component

3G was originally intended to have an integrated satellite component, to provide true worldwide coverage and fill in gaps in the cellular networks. A single satellite/3G handset was sometimes envisaged, surprisingly, since satellite handsets tend to be large.

The classic picture – seemingly compulsory in any description of 3G – is of a layered architecture of radio cells (Figure 2.1). There are mega-cells for satellites, macro-cells for wide-area coverage (rural areas), micro-cells for urban coverage and pico-cells for indoor use. There is a mixture of public and private use and always a satellite hovering somewhere in the background.

In terms of forming this vision of 3G much of the early work was done in the research programmes of the European Community; for example the RACE (Research and Development in Advanced Communications Technologies in Europe) programme with projects such as MONET (looking at the transport and signalling technologies for 3G) and FRAMES (evaluating the candidate

Figure 2-1. Classic 3G layer diagram. (Source: IP for 3G. Reproduced by Permission of © 2002 John Wiley & Sons Ltd).

air interface technologies). In terms of standards, ETSI (European Telecommunications Standards Institute) completed development of GSM phase 2 and this was, at the time, intended to be the final version of GSM and 3G was supposed to totally supersede it and all other 2G systems. As a result European standardisation work on 3G, prior to 1996, was carried out within an ETSI GSM group called, interestingly, SMG5 (Special Mobile Group).

2.3.2 1996–1998 – The IMT 2000 Trimester

It is now appropriate to talk of UMTS – as the developing European concept was being called. In the case of UMTS the Global Multimedia Mobility report (Ref. 4) was endorsed by ETSI and set out the framework for UMTS standardisation. The UMTS Forum – a pressure group of manufacturers and operators produced the influential UMTS forum report (www.umts-forum.org) covering all nonstandardisation aspects UMTS such as regulation, market needs and spectrum requirements. As far as UMTS standardisation was concerned ETSI transferred the standardisation work from SMG5 to the various GSM groups working on the air interface, access radio network and core network.

In Europe there were five different proposals for the air interface – most easily classified by their Medium Access Control (MAC) schemes – in other words how they allowed a number of users to share the same spectrum. Basically there was time division (TDMA – Time Division Multiple Access), frequency division (OFDM- Orthogonal Frequency Division Multiple Access) and code division proposals (CDMA). In January 1998 ETSI chose two variants of CDMA – Wideband CDMA (W-CDMA) and time division CDMA – the latter is basically a hybrid with both time and code being allocated to separate users. W-CDMA was designated to operate in paired spectrum (a band of spectrum for uplink and another (separated) band for downlink) and is referred to as the FDD (Frequency Division Duplex) mode since frequency is used to differentiate between the up and down traffic. In the unpaired spectrum, a single monolithic block of spectrum, the TD-CDMA scheme was designated and this has to use time slots to differentiate between up and down traffic (FDD will not work for unpaired spectrum – see the section on spectrum below for more details), and so is called the TDD (Time Division Duplex) mode of UMTS.

In comparison GSM is a FDD/TDMA system – frequency is used to separate up and down link traffic and time division is used to separate the different mobiles using the same up (or down) frequency.

Part of the reason behind the decision to go with W-CDMA for UMTS was to allow harmonisation with Japanese standardisation.

Unfortunately in North America the situation was more complicated; firstly parts of the 3G designated spectrum had been licensed to 2G operators and other parts used by satellites; secondly the US already had an existing CDMA system called cdmaOne that is used for voice. It was felt that a CDMA system for North America needed to be developed from cdmaOne – with a bit rate that was a multiple of the cdmaOne rate. Consequently the ITU

IMT2000 designation	Common term	Duplex type
IMT-DS Direct Sequence CDMA	W-CDMA	FDD
IMT-MC Multi Carrier CDMA	Cdma2000	FDD
IMT-TD Time Division CDMA	TD-CDMA And TD-SCDMA	TDD
IMT-SC Single Carrier	EOGE	FDD
IMT-FT Frequency Time	DECT	TDD

Table 2-1. The IMT 2000 family of 3G standards (Note : UMTS consists of W-CDMA and TD-CDMA).

recognised a third CDMA system – in addition to the two European systems – called cdma2000. It was also felt that the lack of 3G spectrum necessitated an upgrade route for 2G TDMA systems – resulting in a new TDMA standard – called UMC-136 that is effectively identical to a proposed enhancement to GSM called EDGE (Enhanced Data rates for Global Evolution). Basically this takes advantage of the fact that the signal to noise ratio (and hence potential data capacity) of a TDMA link falls as the mobile moves away from the base-station. Users close to base-stations essentially have such a good link that they can increase their bit rate without incurring errors. By using smaller cells or adapting the rate to the signal to noise ratio then, on average, the bit rate can be increased (much more about EDGE can be found in Chapter 4). In CDMA systems the signal to noise ratio is similar throughout the cell.

Finally the DECT (Digital European Cordless Telecommunications) – developed by ETSI for digital cordless applications and used, for example, in household cordless phones – inhabits the 3G spectrum and has been included as the fifth member of the IMT–2000 family of 3G standards (Table 2.1) as the ITU now called the FPLMTS vision.

During this period 3G progressed from its "martini" vision – "anytime, anyplace, anywhere", to a system much closer, in many respects, to the existing 2G networks. True the air interface was a radical change from TDMA – it promised better spectral efficiency, bandwidth on demand and broadband connections. But the core networks chosen for both UMTS and cdma2000 were based on existing 2G networks; in the case of UMTS an evolved GSM core and for cdma2000 an evolved IS 41 core (another time division circuit switching technology standard). The major reason for this was the desire by the existing 2G operators and manufacturers to reuse as much existing equipment, development effort and services as possible. Another reason was the requirement for GSM to UMTS handover – recognising that UMTS coverage was limited in the early years of roll-out.

The radio access network for UMTS was also new – supporting certain technical requirements of the new CDMA technology and also the resource management for multimedia sessions. The choice of evolved core network for UMTS is probably the key non IP friendly decision that was taken at this time – meaning that UMTS now supports IP, ATM and X25 packets using a common way of wrapping them up and transporting them over an underlying IP network. (X25 is an archaic and heavyweight packet switching technology that pre-dates IP and ATM). In the meantime X25 has become totally defunct as a packet switching technology, ATM has been confined to the network core and IP has become ubiquitous – meaning that IP packets are wrapped up and carried within outer IP packets because of a no longer useful legacy requirement to support X25!

2.3.3 1998 Onwards – The Standardisation Trimester

After 1998 the function of developing and finalising the standards for UMTS and cdma2000 passed to two new standards bodies: 3GPP and 3GPP2 respectively. These bodies completed the first version (or release) of the respective standards (e.g. R3 – formally known as Release 99 for UMTS) in 1999 and these are the standards that most current networks have been built to. New releases of UMTS have continued – we now (Q2 2008) have R7 with R8 being finalised. Gradually, some of the features of these standards have been implemented within 3G networks. For example R5 HSDPA (High Speed Download Packet Access) has been launched in 198 commercial HSDPA networks in 86 countries (Q1 2008 – Ref. 5) and R6 HSUPA (High Speed Upload Packet Access) has 36 commercial HSUPA networks in 27 countries. Chapter 4 covers much more of the 3G story from 2002 onwards when deployment began in earnest.

2.3.4 3G Spectrum – The "Fuel" of Mobile Systems

Now is a good time to consider spectrum allocation decisions as these have a key impact on the 3G vision in terms of the services (e.g. bandwidth or quality) that can be provided and the economics of providing them.

In any cellular system a single transmitter can only cover a finite area before the signal to noise ratio between the mobiles and base-stations becomes too poor for reliable transmission. Neighbouring base-stations must then be set up and the whole area divided into cells on the basis of radio transmission characteristics and traffic density: so long as the neighbouring cells operate on a different frequency (e.g. GSM/D-AMPS) or different spreading code (e.g. W-CDMA or cdmaOne) – see Figure 2.2. Calls are handed-over between cells by arranging for the mobile to use a new frequency, code or time-slot. It is a great, but profitable and very serious, game of simulation and measurement to estimate and optimise the capacity of different transmission technologies. For example it was originally estimated that W-CDMA would offer a ten times improvement in transmission efficiency (in terms of bits transmitted per Hz of spectrum) over TDMA (Time Division Multiple Access – such as GSM and D-AMPS) – in practice this turned out to be about 4–5 times.

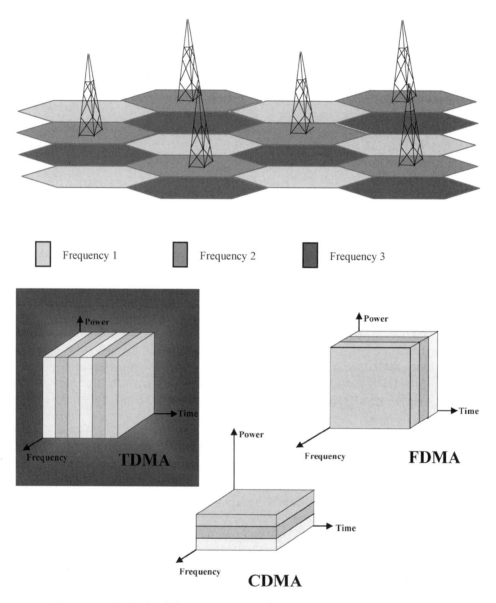

Figure 2-2. *Typical cellular system (top) and (lower) difference between TDMA, FDMA and CDMA. (Source: IP for 3G. Reproduced by permission of © 2002 John Wiley & Sons Ltd.)*

In general terms for voice traffic the capacity of any cellular system is given by:

$$\text{Capacity(users/km}^2\text{)} = \text{Constant} \times \text{spectrum bandwidth available(Hz)}$$
$$\times \text{ Efficiency(bits/s per Hz)} \times \text{density of cells(base-stations/km}^2\text{)/band-width of a call(bit/s)}$$

The constant depends on the precise traffic characteristics – how often users make calls and how long they last as well as how likely they are to move to another base station and the quality desired – the chance of a user failing to make a call because the network is busy or the chance of a call being dropped on hand-over.

Typically figures for a 2G system are:

- Bandwidth of a call – 14 kbit/s (voice)

- Bandwidth available 30 MHz (Orange – UK)

- Efficiency 0.05 (Or frequency re-use factor of 20 – meaning that one in 20 cells can use the same frequency with acceptable interference levels).

Now there are several very obvious conclusions that can be drawn from this simple equation: Firstly, you can achieve any capacity you like simply by building a higher base-station density (although this increases costs). Secondly, the higher the bandwidth per call the lower the capacity – so broadband systems offering 2 Mbit/s to each user need about 150 times the spectrum bandwidth of voice systems to support the same number of users (or will support around 150 times less users) – all other things being equal. Thirdly, any major increase in efficiency – for a given capacity – means that either a smaller density of base-stations or less spectrum is required and, given both are very expensive, this is an important research area. For 3G systems, as mentioned above, this factor has improved by 4–5 over current GSM systems. Finally if the bandwidth of a voice call can be halved then the capacity of the system can be doubled; this is the basis of the (now defunct) half-rate (7 kbit/s) voice coding in GSM.

So, given this analysis, it is hard to escape the conclusion that mobile broadband Internet systems, serving large web pages and with video content, need a lot of spectrum. However, radio spectrum is a scarce resource – to operate a cellular mobile system only certain frequencies are feasible: at higher frequencies radio propagation characteristics mean that the cells become smaller and costs rise: For example 900 MHz GSM operators (e.g. Cellnet in the UK) require about half the density of stations – in rural areas – compared to 1800 MHz GSM operators like Orange. The difficulties of finding new spectrum in the 500–3000 MHz range should not be under-emphasised – see Ref. 6 for a lengthy account of the minutiae involved – but, in short, all sorts of military, satellite, private radio and navigation systems and so forth all occupy different parts of the spectrum in different countries. Making progress to reclaim – or "re-farm" as it is known – the spectrum is painfully slow on a global scale. The spectrum bands earmarked for FPLMTS at the World Radio Conference in 1992 were 1885–2025 MHz and 2110–2200 MHz – a total of 230 MHz. However, a number of factors and spectrum management decisions have since eroded this allocation in practice:

- Mobile satellite bands consume 2×30 MHz.

- In the US licences for much of the FPLMTS band have already been sold off for 2G systems.

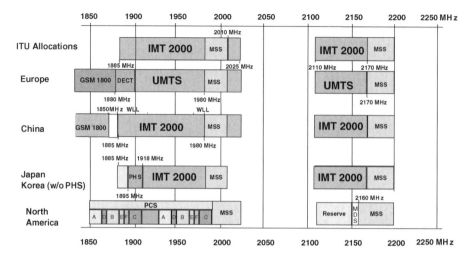

Figure 2-3. Global spectrum allocations for 3G (MSS bands are satellite spectrum). (*Source: IP for 3G.* Reproduced by permission of © 2002 John Wiley & Sons Ltd.)

- Part of the bands (1885–1900 MHz) overlap with the European DECT system.

- The FPLMTS bands are generally asymmetrical (preventing paired spectrum allocations – see below).

All of which means that only 2 × 60 MHz and an odd 15 MHz of unpaired spectrum was available for 3G in Europe and much less in the US. Paired spectrum is important – it means equal chunks of spectrum separated by a gap – one part is used for up-link communications and the other for down-link transmission. Without the gap separating them up and down-link transmissions would interfere at the base-station and mobile if they transmitted and received simultaneously. By comparison, in the UK today, 2 × 100 MHz is available for GSM, shared by four operators. Figure 2.3 shows the general world position on 3G spectrum.

In the UK auction/licensing process there were a dozen or so bidders chasing five licences resulting in three each getting 2*10 MHz and two buying 2*15 MHz of paired spectrum per operator. The five successful bidders paid £22 billion – yes billion – for the spectrum (only the German auction raised similar sums). So far the unpaired spectrum has been unused and operators have concentrated on building UMTS coverage.

More spectrum is expected to be available from re-farming existing GSM wavelengths. In the UK, for example, there are two GSM operators each with: 2 × 17.5 MHz of 900 MHz plus 2*10 MHz of 1800 MHz spectrum. Two other operators have 2 × 30 MHz of 1800 MHz GSM spectrum. Although the operators see the GSM network continuing to 2018 (Ref. 7) it is expected that more voice users will move to UMTS and allow some of this spectrum to be re-farmed. Re-farming isn't as easy as planting potatoes, however. Firstly,

	Europe, Middle East, Africa	Asia	North America
New spectrum available now	2.6 GHz (Japan)	1.5 GHz (Japan)	700 MHz
		2.6 GHz (Japan)	
		2.1 GHz (Japan)	
		2.3–2.4 GHz (China)	
Future new spectrum	900 MHz (re-farm)		850 MHz (re-farm)
	1.8 GHz (re-farm)	1.8 GHz (re-farm)	1.9 GHz (re-farm)
	450 MHz (re-farm)	470–854 MHz (digital dividend) Future	
	470–854 MHz (digital dividend)		

Table 2-2. New spectrum likely to be available for licenced mobile systems.

there are difficult questions about how much will be re-farmed and who will get it – some of the original owners were given the spectrum for free and it is expected that national regulators will get involved.

Further spectrum is being sold around the world – at 700 MHz in the US, 1.5 GHz in Japan, 2.3–2.4 GHz in China and at 2.6 GHz in Europe, Africa, the Middle East and Asia Pacific. The 2.6 GHz auction in Europe, for example, could offer up to three lots of 2*20 MHz FDD plus one lot of 2*10 MHz FDD and some 50 MHz of TDD spectrum. This so called UMTS extension band (2.5 to 2.69 GHz) is mostly being auctioned on a "technology neutral" basis – meaning any mobile system can be used – and the exact packages of spectrum, along with the price, varies from country to country. At the time of writing the Swedish and Norwegian auction had just finished and the UK auction was scheduled for January 2009. The larger allocations (2*20 MHz) are significant as they are much more efficient when used with new technologies (such as LTE, described in Chapter 4).

There is also the so-called digital dividend – the result of switching analogue TV signals to digital. Typically analogue TV is broadcast on frequencies from 400 MHz and 850 MHz. Much of this is expected to be available for a variety of uses – including mobile systems – in the next three to five years in much of Europe and Asia (Table 2.2).

2.4 UMTS Network Overview

In order to illustrate the operation of a UMTS network we will now describe a day in the life of a typical UMTS user – this sort of illustration is often called a usage case or a scenario. We will introduce the major network elements – the base-stations and switches etc. – as well as the functionally that they provide. This at least has the merit of avoiding a very sterile list of the network elements and serves as a high level guide to the detailed description of UMTS functionality that follows. Figure 2.4 shows the network for the scenario.

Mary Jones, who is 19 years old, just arrived at the technical Polytechnic of Darmstadt. She is lucky that her doting father has decided to equip her with a 3G terminal for when she lives away from home.

Mary first turns her terminal on before breakfast and is asked to enter her personal PIN code.[1] This actually authenticates her to the USIM (UMTS Subscriber Identity Module) – a smart card that is present within her terminal. The terminal then searches for a network, obtains synchronisation with a local base-station and, after listening to the information on the cell's broadcast channel, attempts to attach to the network. Mary's subscription to T-Nova is based on a 15-digit number (which is not her telephone number) identifying the USIM inside her terminal. This number is sent by the network to a large database – called the home location register (HLR) located in the T-Nova core network. Both the HLR and Mary's USIM share a 128bit secret key – this is applied by the HLR to a random number using a one-way mathematical function (one that is easy to compute but very hard to invert). The result and the random number are sent to the network which challenges Mary's USIM with the random number and accepts her only if it replies with the same result as that sent from the HLR.

After attaching to the network Mary decides to call her dad – perhaps, although unlikely, to thank him for the 3G terminal! The UMTS core network is divided into two halves – one half dealing with circuit switched (constant bit rate) calls – called the circuit switched domain – and the other – the packet-switched domain – routing packets sessions. At this time Mary attempts to make a voice call and her terminal utilises the connection management functions of UMTS. Firstly, the terminal signals to the circuit switch that it requires a circuit connection to a particular number – this switch is an MSC (mobile switching centre). The MSC has previously downloaded data from the HLR when Mary signed on, into a local database called the visitor location register (VLR) and so knows if she is permitted to call this number: e.g. she may be barred from international calls. If the call is possible the switch sets up the resources needed in both the core and radio access networks. This involves checking whether circuits are available at the MSC and also whether the radio access network has the resources to support the call. Assuming the call is allowed and resources are available a constant bit rate connection is set up from the terminal, over the air interface and across the radio access network to the MSC – for mobile voice this will typically be 10kbit/s or so. Assuming Mary's dad is located on the public fixed network then the MSC transcodes the speech to a fill a 64kbit/s speech circuit (the normal connection for fixed network voice) and transports this to a

---

[1] OK so 19-year-olds never turn the phone off, even at night, but how else can we cover network discovery and sign-on!

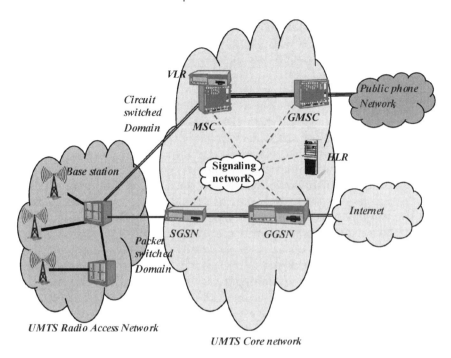

Figure 2-4. UMTS Architecture (R99/R3). (Source: IP for 3G. Reproduced by permission of © 2002 John Wiley & Sons Ltd.)

gateway switch (the gateway MSC – GMSC) to be switched into the public fixed telephone network.

When the call ends both the MSC and GMSC are involved in producing Call Detail Records (CDR) – with such information as: called and calling party identity; resources used; timestamps and element identity. The CDRs are forwarded to a billing server where the appropriate entry is made on Mary's billing record.

Mary leaves her terminal powered on – so that it moves from being Mobility Management (MM) connected to being MM-idle (when it was turned completely off it was MM-detached). Mary then boards a bus for the Polytechnic and passes the radio coverage of a number of UMTS base-stations. In order to avoid excessive location update messages from the terminal the system groups large numbers of cells into a location area. The location area identifier is broadcast by the cells in the information they broadcast to all terminals. If Mary's terminal crosses into a new location area then a location update message is sent to the MSC and is also stored in the HLR.

When Tom tries to call Mary – he is ringing from another mobile network – his connection control messages are received by the T-Nova GMSC. The GMSC does a look-up in the HLR, using the dialled number (i.e. Mary's telephone number) as a key – this gives her current serving MSC and location area and the call set up request is forwarded to the serving MSC. Mary's terminal is then paged within the location area – in other words all the cells in that area request Mary's terminal to identify the cell that it is currently in. The terminal can happily sit in the MM-idle state, listening to the broadcast messages and doing occasional location area updates without expending very much energy.

Mary and Tom begin a conversation but as Mary is still on the bus, the network needs to handover the connection from one base-station to another as she travels along. In CDMA systems, however, terminals are often connected to several cells at once, especially during handover – receiving multiple copies of the same bits of information and combining them to produce a much lower error rate than would be the case for a single radio connection. When the handover is achieved by having simultaneous connections to more than one base-station it is called soft-handover and in UMTS the base-stations connected to the mobile are known as the active set.

Mary attends her first lecture of the day on relativity and gets a little lost on the concept of time dilation – she decides to browse the Internet for some extra information. Before starting a browsing session her terminal is in the PMM (Packet Mobility Management) idle state – in order to send or receive packets the terminal must create what is called a PDP (packet data protocol) context. A PDP context basically signals to the SGSN and GGSN (Serving GPRS Support Node and Gateway GPRS Support Node) – which are the packet domain equivalent of the MSC and GMC switches – to set up the context for a packet transfer session. What this means is that Mary's terminal acquires an IP address, the GSNs are aware of the Quality of Service (QoS) requested for the packet session and that they have set up some parts of the packet transfer path across the core network in advance. Possible QoS classes for packet transfer, with typical application that might use them, are: Conversational (e.g. voice), streaming (e.g. streamed video), interactive (e.g. web browsing) and background (file transfer). (All circuit switched connections are conversational). Once Mary has set up a PDP context the Session Management (SM) state of her terminal moves from inactive to active.

When Mary actually begins browsing her terminal sends a request for resources to send the IP packet(s) and, if the air interface, radio access and core networks have sufficient resources to transfer the packet within the QoS constraints of the interactive class then the terminal is signalled to transmit the packets. Mary is able to find some useful material and eventually stops browsing and deactivates her PDP context when she closes the browser application.

During the afternoon lecture Mary has her 3G terminal set to divert incoming voice calls to her mail box. Tom tries to ring her and is frustrated by the voice mail – having some really important news about a party that evening. He sends her an Instant Message. When this message is received by the T-Nova gateway it is able to look in the HLR and determine that Mary is attached to the network but has no PDP context active – it also only knows her location for packet services within the accuracy of a Routing Area (RA). This is completely analogous to the circuit switched case and a paging message is broadcast requesting Mary's terminal to set up a PDP context so that the urgent IM can be transferred.[2]

In this scenario we have briefly looked at the elements within the UMTS network and how they provide the basic functions of: security, connection management, QoS, mobility management and transport of bits for both the circuit and packet switched domains. In the next section we go into some of the grisly details and expand on some of these points.

We have not yet said much about the role of the Radio Access Network and the air interface. The Radio Access Network stretches from the base-station, through a node called the Radio Network Controller, to the SGSN/MSC. The RAN is responsible for mobility management – nearly all terminal mobility is hidden from the core network being managed by the RAN. The RAN is also responsible for allocating the resources across the air interface and within the RAN to support the requested QoS.

2.5 UMTS Network Details

To avoid a very lengthy description we will mostly follow UMTS (Universal Mobile Telecommunications System) which is the European/Japanese member of the IMT-2000 family.

It is convenient to break 3G networks into an architecture, what the building blocks (switching centres, gateways) are and how they are connected (interfaces), and four functions that are distributed across the architecture:

- Transport – how the bits are routed/switched around the network.

- Security – how users are identified, authorised and billed.

- Quality of Service – how users get better than best effort service.

- Mobility management – the tracking of users and hand-over of calls between cells.

You could easily break down the PSTN in this way – mobility management would be reduced to a cordless phone! However the building blocks would be the terminal, local exchange and main switching centre. The bits would be transported by 64kbit/s switching technology from the exchange level and quality would be provided by provisioning using Erlang's formula – and you either get 64kbit/s or nothing! Finally phones are identified by an E164 number (01473....) and being named on the contract with the phone company makes you responsible for all call charges – the phone is secured by your locked front door!

---

[2] Of course he might just send an SMS – but IM is an Internet service and (in theory) faster than SMS.

2.5.1 UMTS Architecture – Introducing the Major Network Elements and their Relationships

UMTS is divided into three major parts: the air interface, the UMTS Terrestrial Radio Access Network (UTRAN) and the core network. The first release of the UMTS network (Figure 2.5) – R3, the Release previously known as R99 – consists of an enhanced GSM phase 2 core network (CN) and a wholly new Radio Access Network (called the UMTS Terrestrial Radio Access Network or UTRAN).

For readers familiar with GSM the MSC, G-MSC, HLR and VLR (See Ref. 8 for more details on GSM) are simply the normal GSM components but with added 3G functionality. The UMTS RNC (Radio Network Controller) can be considered to be roughly the equivalent of the Base Station Controller (BSC) in GSM and the Node Bs equate approximately to the GSM base-stations (BTS – Base Transceiver Station).

The RNCs and base-stations are collectively known as the UTRAN (UMTS Terrestrial Radio Access Network). From the UTRAN to the Core the network is divided into packet and circuit switched parts: The Interface between the radio access and core network (I_u) being in effect two interfaces: I_u (PS – Packet Switched) and I_u (CS – Circuit Switched). Packet traffic is concentrated in a new switching element – the SGSN (Serving GPRS Support Node). The boundary of

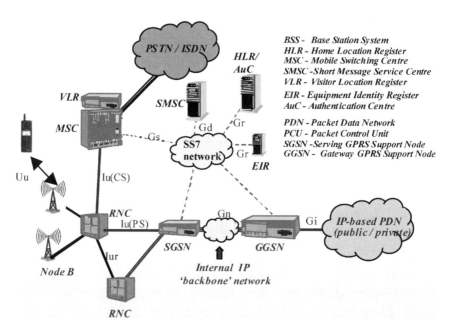

Figure 2-5. UMTS R3 (Release 99) Architecture (Detail). (Source: IP for 3G. Reproduced by permission of © 2002 John Wiley & Sons Ltd.)

the UMTS core network for packets is the GGSN (Gateway GPRS Support Node) which is very like a normal IP gateway and connects to corporate intranets or the Internet.

Below is a quick guide to some of the functionality of each of these elements and interfaces:

- 3G Base-station (Node B)

  The base-station is mainly responsible for the conversion and transmission/reception of data on the U_u interface (Figure 2.5) to the mobile. It performs error correction, rate adaptation, modulation and spreading on the air interface. Each Node B may have a number of radio transmitters and cover a number of sectors. A typical 3G base-station would have one or three sectors depending on the environments (urban, rural etc.) and the traffic level. In the case of the three sectors each would have a different scrambling code (which is described later in the chapter). Each sector can also have a number of 5 MHz carriers – in the UK the MNOs have 2 or 3 5 MHz carriers – depending on the traffic level. As we will see in Chapter 4 the second and third carrier are typically being used for mobile data and the first carrier for voice – although they can be mixed on a carrier. The Node B can also achieve soft handover between its own transmitters and also send measurement reports to the RNC: this is called softer handover.

- RNC

  The RNC is an ATM switch which can multiplex/demultiplex user packet and circuit data together. Unlike in GSM, RNCs are connected together (through the I_{ur} interface) and can handle all radio resourcing issues autonomously. Each RNC controls a number of Node Bs – the whole lot being known as an RNS – Radio Network System (thought you'd have got the hang of the TLAs – Three Letter Abbreviations! by now!). The RNC controls congestion and soft hand-over (involving different Node Bs) as well as being responsible for operation and maintenance (monitoring, performance data, alarms and so forth) within the RNS.

- SGSN

  The SGSN is responsible for session management, producing charging information, and lawful interception, it also routes packets to the correct RNC. Functions such as attach/detach, setting up of sessions and establishing QoS paths for them are handled by the SGSN.

- GGSN

  A GGSN is rather like an IP gateway and border router – it contains a firewall, has methods of allocating IP addresses and can forward requests for service to corporate Intranets (as in Internet/Intranet connections today). GGSNs also produce charging records.

- MSC

  The Mobile Switching Centre/Visitor Location Register handles connection orientated circuit switching responsibilities including connection management (setting up the circuits) and mobility management tasks (e.g. location registration and paging). It is also responsible for some security functions and Call Detail Record (CDR) generation for billing purposes.

- GMSC

  The Gateway MSC deals with incoming and outgoing connections to external networks (such as the public fixed telephony network) for circuit switched traffic. For incoming calls it looks up the serving MSC within the HLR and sets up the connection to the MSC.

- HLR

  The home location register familiar from GSM is just a big data base with information about users, their services (e.g. are they pre or post pay? Is roaming switched on? What QoS classes they have subscribed to?). Obviously new fields have been added for UMTS – especially relating to data services.

2.5.2 UMTS Security

Security in a mobile network covers a wide range of possible issues affecting the supply of and payment for services. Typical security threats and issues might be:

- Authentication – is the person obtaining service the person who is paying for it?

- Confidentiality of data – is anyone eavesdropping on my data/conversations?

- Confidentiality of location – can anybody discover my location without authorisation?

- Denial of service – can anybody deny me service (e.g. sending false update messages about my terminal location) to prevent me getting some service. An example of this might be when I am bidding in an auction and other bidders wish to prevent me continuing to bid against them.

- Impersonation – can users take my mobile identity – and get free service? Or access to my information? Can sophisticated criminals set up false base-stations that collect information about users or their data?

   In UMTS there are four main ways in which threats and issues like these are addressed:

- Mutual authentication between the user and the network.

- Signalling integrity protection within the RAN.

- Encryption of user data in the RAN and over the air interface.

- Use of temporary identifiers.

Mutual authentication – of the user to the network and of the network to the user is based around the USIM (UMTS Subscriber Identity Module). This is a smart card (i.e. one with memory and a processor in it) and each USIM is identified by a (different) 15 digit number – the International Mobile Subscriber Identity (IMSI) – Note the ISMI is separate from the phone number (07702 XXXXXX say), which is known as the Mobile ISDN number and can be changed (e.g. in number portability you can move to a new MNO (and get a new IMSI) whilst keeping your mobile number). When you switch on a signalling message is sent to the HLR (your home HLR if you are roaming on a foreign network – identified by your IMSI) containing your IMSI and the "address" of MSC that you are registering with. The HLR (actually in a sub part of the HLA called the authentication centre AuC) generates a random number (RAND) and computes the result of applying a one-way mathematical procedure (XRES), which involves a 128 bit secret key known only to the SIM and the HLR, to RAND. The one way function is very difficult to invert – knowledge of the random number and the result of the function does not allow the key to be easily found. The HLR sends this result and random number to the visited-MSC– which challenges the USIM with the random number and compares the result with that supplied by the HLR. If they match the USIM is authenticated. The MSC can download a whole range of keys to store for future use (in the VLR)– which is why when you first turn on your mobile abroad it seems to take ages to register but, subsequently, is much quicker to attach. Note that at no time does the secret key leave the SIM or HLR – there are no confirmed cases of hackers gaining access to these keys.

A second feature of UMTS is that it allows the user to authenticate the network – to guard against the possibility of "false" base-stations (i.e. like bogus bank machines that villains use to collect data to make illegal cards). When the home network HLR receives the authentication request from the serving network MSC it actually uses the secret key to generate three more numbers – known as AUTN, CK and IK. The set (XRES, AUTN, CK and IK) are known as the authentication vectors (Figure 2.6)

Figure 2-6. UMTS authentication. (Source: IP for 3G. Reproduced by permission of © 2002 John Wiley & Sons Ltd.)

Both HLR and USIM also keep a sequence number (SQN) of messages exchanged that is not revealed to the network. The MSC sends RAND and AUTN to the USIM that is then able to calculate the RES, SQN CK and IK. It sends RES to the network for comparison with XRES – to authenticate itself – but also checks that the computed value of the sequence number with its own version to authenticate the network to itself.

Another feature introduced is an integrity key (IK) – distributed to the mobile and a network by the HLR, as described above, so that they can mutually authenticate signalling messages. This takes care of the sort of situation where false information might be sent to the network or to the mobile. This would cover the auction example where a rival bidder sends a false signal that I want to detach or have moved to a new base-station just as we are reaching the end of a bidding session.

In addition to the challenge/response the HLR generates a cipher key (CK) and distributes this to the MSC and USIM. The cipher key is used to encrypt the user data over the air from the terminal to the RNC and is passed to RNC by the MSC when a connection or session is set up.

UMTS allows the terminal to encrypt its IMSI at first connection to the network by using a group key – it sends the MSC/SGSN the coded IMSI and the group name that is then used by the HLR to apply the appropriate group key. The IMSI is actually only sent over the air at registration or when the network gets lost to prevent the capture of UMTS identities. After first registration the terminal is identified by a Temporary Mobile Subscriber Identifier (TMSI) for the circuit switched domain and a Packet Temporary Mobile Subscriber Identifier (P-TMSI). These temporary identifiers – and the encryption of the IMSI at first attach should prevent IMSI being captured for malicious use and impersonation of users.

One, final, level of security is performed on the mobile equipment itself, as opposed to the mobile subscriber (for example putting your SIM in someone else's phone doesn't always work!)

Each terminal is identified by a unique International Mobile Equipment Identity (IMEI) number and a list of IMEIs in the network is stored in the Equipment Identity Register (EIR). An IMEI query to the EIR is sent at each registration and returns one of the following:

- White-listed
  The terminal is allowed to connect to the network.

- Grey-listed
  The terminal is under observation from the network.

- Black-listed
  The terminal has either been reported stolen, or is not type approved (wrong type of terminal). Connection to be refused.

Good references for UMTS security are given in Ref. 9.

2.5.3 UMTS Communication Management

2.5.3.1 Connection Management

For the circuit switched domain the connection management function is carried out in the MSC and GMC. Connection management is responsible for number analysis (can the user make an international call), routing (setting up a circuit to the appropriate GMSC for the call and charging (i.e. generation of Call Detail Records). The MSC is also responsible for the transcoding of low bit rate mobile voice (10 kbit/s or so – in UMTS the voice data rate is variable) into 64kbit/s streams that are standard in the fixed telephony world.

The GMSC is responsible for the actual connection to other circuit-based networks and also for any translation of signalling messages that is required.

2.5.3.2 Session Management

In the packet domain the user needs to set up a PDP context (Packet Data Protocol Context) in order to send or receive any packets. The PDP context describes the connection to the external packet data network (e.g. the Internet): is it IP? What is the network called? (e.g. BT Corporate network) What quality do I want for this connection (delay, loss) and how much bandwidth do I want (QoS Profile).

The steps involved in setting up a PDP context are as follows (Figure 2.7):

1. The terminal requests PDP context activation.

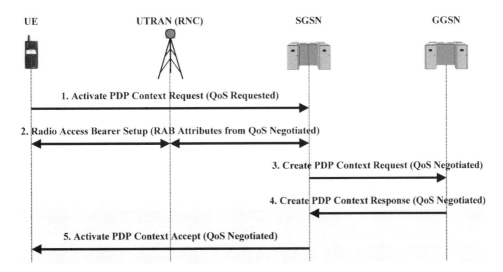

Figure 2-7. PDP context set up. (Source: IP for 3G. Reproduced by permission of © 2002 John Wiley & Sons Ltd.)

2. The SGSN checks the request against subscription information received from the HLR (during the attachment). If the requested QoS is not included in the subscription then it may be rejected/re-negotiated.

3. The Access Point Name (name of external network) is sent, by the SGSN, to a DNS server (IP Domain Name Server – normal Internet-style name->IP address look up to find the IP address of the GGSN that is connected to the required network).

4. The SGSN tries to set up the radio access bearers – this can result in re-negotiation of QoS.

5. The SGSN sends a PDP create context message to the GGSN and this may be accepted or declined (e.g. if the GGSN is overloaded).

6. An IP tunnel is set up between the SGSN and the relevant GGSN – with a tunnel ID (this will be explained in the next section).

7. An PDP address is assigned to the mobile.

8. The PDP context is stored in the: mobile, SGSN, GGSN and HLR.

The PDP address will in practice be an IP address (Although UMTS can carry X25 and PPP – point to point protocol packets as well) and this can be either static or dynamically assigned. In static addressing the mobile always has the same IP address – perhaps because it is connecting to a corporate network whose security requires an address from the corporate range.

In dynamic allocation the address can come from a pool held by the GGSN and allocated by DHCP (Dynamic Host Configuration Protocol – again normal Internet-style IP address allocation) or from a remote corporate or ISP network. The GGSN includes a RADIUS client that can forward password and authentication messages to external servers (as happens in internet access today). This would typically be the case where users are connecting to their corporate networks. Note that most MNOs do not offer a public IP address for mobile data but rather use NAT (Network Address Translation) and offer only private addresses – as do most fixed ISPs.

UMTS also contains the concept of a secondary PDP context (also called a multiple PDP context – Figure 2.8). In GPRS if you want to run two different applications, with different QoS requirements – such as video streaming and www browsing – then you need two different PDP contexts and, consequently, two different PDP (i.e. IP) addresses. In UMTS R99 the secondary PDP context concept allows multiple application flows to use the same PDP type, address and Access Point Name (i.e. external network) but with different QoS profiles. The flows are differentiated by an NSAPI (Network layer Service Access Point Identifier – a number from 0 to 15). We will look at the mapping of the various identifiers and addresses later in the mobility management section.

A traffic flow template (TFT) is used to direct packets addressed to the same PDP address to different secondary PDP contexts. For example if our user is browsing and wants to watch a movie clip – a long one so she wants to stream it

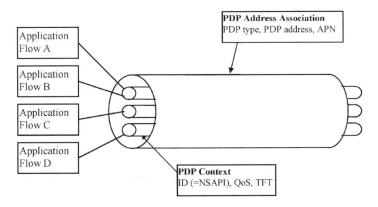

Figure 2-8. Multiple PDP contexts. (Source: IP for 3G. Reproduced by permission of © 2002 John Wiley & Sons Ltd.)

rather than download it – then the browser might activate a secondary PDP context suitable for video streaming. When the video and HTTP packets arrive at the GGSN they all have the same destination IP address (PDP address). The packet flow template allows other aspects (source address, port number, flow label. . .) to be used to assign them to the correct context and hence, QoS. In this case the source address (or source address and source port number) might be used to differentiate between the flows.

A PDP context will only remain active for a certain length of time after the last packet transmission. In other words I might set up a PDP context to browse some web pages and then stop using the terminal. Obviously I am tying up network resources (e.g. IP addresses) – and I am almost certainly not paying for them (if I pay per packet or by subscription). The network, therefore, deactivates the PDP after a suitable time. Some can be long lived in the absence of traffic if the mobile operator selects this option which results in much quicker set up times – e.g. when I browse a few pages stop to read something and then click on a link the delay can be quite long if the PDP context has to be reset. It might seem from this that UMTS packet users are confined to user-initiated sessions (the equivalent of outgoing calls only) – but there also exists a mechanism to request users to set up a PDP context. This might be when users have a fixed IP address – so that the GGSN can accept an incoming instant message (say) and use the IP address as a key in the HLR and obtain the address of the SGSN which the mobile is associated with. When the mobile attached it initially joined to an SGSN and the address of that SGSN was recorded in the HLR – as were subsequent movements of the mobile into regions (routing areas) controlled by other SGSNs. The SGSN can send a PDP set up request to the mobile. Of course the GGSN has to be careful not to request a PDP context every time a piece of junk email is received! The facility will be more useful when Session Initiation Protocol is used widely for peer-to-peer session initiation (see Chapter 6 for more on this).

2.5.4 UMTS QoS

We saw earlier that when users set up PDP contexts they included a QoS profile – in this section we look at how QoS is described within a UMTS network.

UMTS contains the concept of layered QoS – so that a particular bearer service uses the services of the layer below (Figure 2.9). What does this mean? Well a bearer is a term for a QoS guaranteed circuit or QoS treatment of packets. A concrete example would be that packets leaving the UTRAN – on the Iu (PS) interface – are carried on ATM virtual circuits (that give guaranteed QoS). So the CN (Core Network) bearer might be an ATM network with virtual circuits offering different QoS characteristics.

Both the local and external bearers are not part of UMTS – but obviously impact on the end to end QoS. The local bearer might be a Bluetooth link from a 3G mobile phone to a laptop say. In a similar way, the external bearer might, for example, be a DiffServ network operated by an ISP (refer to the QoS chapter for more details).

At the UMTS bearer level, where PDP contexts are created, all UMTS packet services are deemed to fall into one of four classes (Table 2.3) – basically classified by their real-time needs: i.e. the delay they will tolerate.

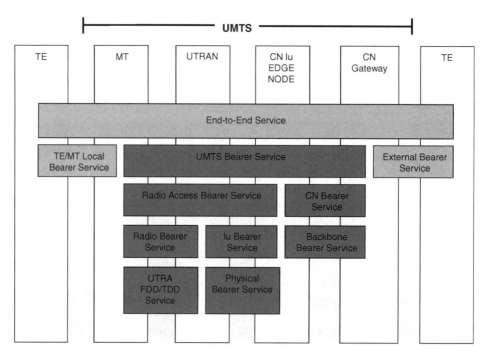

Figure 2-9. UMTS QoS architecture. (Source: IP for 3G. Reproduced by permission of © 2002 John Wiley & Sons Ltd).

Traffic class	Conversational class	Streaming class	Interactive class	Background
	conversational RT	streaming RT	Interactive best effort	Background best effort
Delay	$\ll$ 1 sec	<10 sec	Approx. 1 sec	>10 sec
Example: Error tolerant	Conversational voice & video	Streaming audio & video	Voice messaging	Fax
Example: Error intolerant	telnet, interactive games	FTP, still image, paging	e-commerce, www browsing	email arrival notification
Fundamental characteristics	Preserve time relation (variation) between information entities of the stream	Preserve time relation (variation) between information entities of the stream	Request response pattern	Destination is not expecting the data within a certain time
	Conversational pattern (stringent and low delay)		Preserve payload content	Preserve payload content

Table 2-3. UMTS traffic classes.

Conversational and streaming classes are intended for time-sensitive flows – conversational for delay sensitive traffic such as VoIP (voice over IP). In the case of streaming traffic – such as watching a video broadcast say – much larger buffering is possible and so delays can be relaxed and greater error protection provided by error correction techniques that repeat lost packet fragments but add to delays. Interactive and background classes are for bursty, Internet-style, traffic.

When requesting QoS, users invoke a QoS profile that uses the traffic class and seven other parameters to define the requested QoS:

- Maximum bit rate. The maximum bit rate defines the absolute maximum that the network will provide – packets in excess of this rate are liable to being dropped – this is equivalent to the conventional peak rate description and is only supported when resources are available.

- Delivery order. The delivery order specifies if in sequence delivery of SDUs is required (for SDU – Device Data Unit read IP packet)

- Transfer delay.

- Guaranteed bit rate. Only the guaranteed rate is always available at all times and this only applies to the conversational and streaming classes.

- SDU (Service Data Unit) size information. The maximum SDU size.

- Reliability – whether erroneous SDU should be delivered.

- Traffic handling priority. Traffic handling priority is only used within the interactive class to provide multiple QoS sub-levels.

- Allocation/retention policy. Related to the priority of the traffic (this is explained in detail in the UTRAN section later).

There are only certain values allowed for each parameter – more details can be found in the references at the end of the chapter. In practice, however, UMTS QoS has proved to be not that useful. Let me explain. Pre-empting Chapter 4 (which gives real life measurements on 3G networks) the delay in 3G packet networks has been too large (200–300 ms) to allow VoIP. Also VoIP is a bit pointless when you have a CS network fully optimised to handle voice and with 4–5x the capacity of GSM. The public didn't want video services[3] and so the need to support real-time services on the packet network has not yet materialised. For non real-time services – email, web browsing and file transfer mostly – there has not really been any need for QoS. When the networks first rolled out there was lots of spare capacity. Where demand for mobile data has been high – e.g. in central London – congestion on 3G

---

[3] Or, not up to the present time in any case. Whether they will ever take off – after 20 years of products and trials on fixed and mobile phones is a bit like waiting for:IPv6, nuclear fusion and Godot!

packets networks has been occurring. No QoS separation is currently used for mobile data – although there is strict separation from voice traffic. The problem is being solved by rolling out more 3G cells or adding more sectors to existing cells or "lighting up" extra carriers (i.e. chunks of 5 MHz spectrum – 3G MNOs typically have one to four carriers). QoS is useful when the network load is about 70–200% of capacity. Below 70% you don't need QoS as it all fits on anyway. Over 200% you need more capacity and QoS schemes are insufficient (assuming IP bursty traffic and Internet apps). QoS schemes at 200% only work by admission control – i.e. not letting new users join the network. The problem is that having paid the network operator for mobile broadband access the customers get upset if they can't connect and the operators prefer to admit them and let all users suffer congestion. In fixed broadband DSL there are, typically, 30–50 users sharing a 2Mbit/s connection – some ISPs (such as BT) have just started to introduce QoS for downstream traffic with DiffServ markings used to protect voice and (paid for) IPTV traffic. This may well be a development that is copied in the mobile sphere when traffic levels rise.

2.5.5 UMTS Mobility Management

Most of the mobility management in a UMTS system takes place within the RAN; this was actually one of the design goals of the RAN – isolating the core network from mobility events. Nearly all handovers fall into this category and have been duly relegated to the next section about the UTRAN.

At the core network level there are three mobility management states that the terminal can exist in – detached (i.e. switched off), connected and idle – the last two states having a different meaning in the circuit and packet domains. In the circuit switched domain the terminal is always associated with an MSC and the serving MSC's identity is recorded in the HLR. When a terminal has been idle for circuit switched traffic for a given time the network stops tracking it at the cell level and the terminal simply listens to the broadcast channel of the cells. As it roams about the terminal is in the circuit switched mobility management idle mode (MM-idle). Only when it enters a new location area – consisting of a large number of cells – does it inform the network of a change of location. When the user wishes to make a call it performs a procedure called a location update that provides the network with its position at the cell level of detail. Similarly if an incoming call is received for the terminal then the MSC broadcasts a paging request for that terminal which immediately responds with a location update – bringing it into the MM-connected state.

Likewise for the packet mobility management (PMM) – when the terminal hasn't sent or received any packets for long time it ceases to have a PDP context set up and moves to the PMM-idle mode. When a new PDP context is set up – either as a result of the user wanting to send data or a PDP context set up request message – then the terminal moves to the PMM-connected state. When a terminal is in the PMM-idle state it simply listens to broadcast messages and updates the network whenever it passes into a new routing area. Routing areas are actually subsets of location areas but still comprise many cells.

2.5.6 UMTS Core Network Transport

In this section we will look at how data is transported across the core network and how QoS can be achieved. Figure 2.10 shows the user plane protocols for the core and access networks for packet switched traffic.

From the terminal to the RNC, IP packets are carried in PDCP packets. PDCP is Packet Data Convergence Protocol and provides either an acknowledged/ unacknowledged or transparent transfer service. This choice is related to the (backward) error correction that the underlying RLC (Radio Link Control) layer applies – more details of the functions of RLC can be found in the UTRAN section below. Transparent means no error correction is applied at layer 2. The unacknowledged mode detects duplicate and erroneous packets but simply discards them whereas in acknowledged mode the RLC resends missing frames (at layer 2 packets are usually called frames! – e.g. Ethernet frames). The choice of mode is based on the required QoS – re-sending lost or errored frames causes delay and so the acknowledged mode is only used for applications that are delay sensitive. PDCP also performs a compression/decompression function – such as compressing TCP/IP headers.

From the RNC to the SGSN IP packets are tunnelled using a tunnelling protocol called GTP – GPRS tunnelling protocol (Figure 2.11). Another GTP tunnel then runs from the SGSN to the GGSN – allowing hierarchical mobility (SGSN changes will not happen often) as well as lawful interception (phone tapping) at the SGSN.

A tunnelling protocol consists of two pieces of software that take packets and wrap them within new packets such that the entire original packet – including

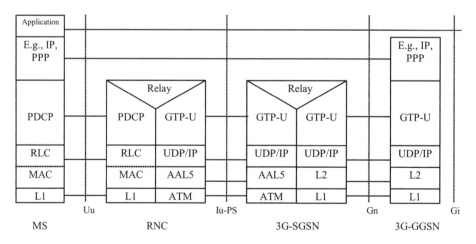

Figure 2-10. UMTS user plane protocols. (Source: IP for 3G. Reproduced by permission of © 2002 John Wiley & Sons Ltd).

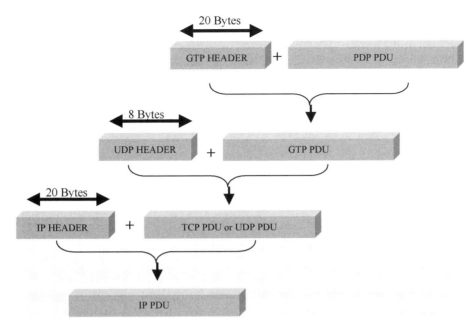

Figure 2-11. GTP-Tunnelling. (Source: IP for 3G. Reproduced by permission of © 2002 John Wiley & Sons Ltd.)

the header becomes the new payload: the original header is not used for routing/switching and is not read whilst encapsulated. A very good analogy is that if I send my friend a letter to his home address his mum puts it in a new envelope, addressed to his college address, and pops it back in the post.

Using GPRS tunnelling protocol UMTS can carry a number of different packets (such as IPv4, IPv6, PPP and X25) over a common infrastructure. GTP packets are formed by adding a header to the underlying PDP packet – the format of this header is shown in Figure 2.12. After forming a GTP packet it is

Bits

Octets	8	7	6	5	4	3	2	1
1		Version		PT	(*)	E	S	PN
2	Message Type							
3	Length (1st Octet)							
4	Length (2nd Octet)							
5	Tunnel Endpoint Identifier (1st Octet)							
6	Tunnel Endpoint Identifier (2nd Octet)							
7	Tunnel Endpoint Identifier (3rd Octet)							
8	Tunnel Endpoint Identifier (4th Octet)							
9	Seqeuence Number (1st Octet)[1) 4)]							
10	Seqeuence Number (2nd Octet)[1) 4)]							
11	N-PDU Num ber[2) 4)]							
12	Next Extensio n Header Type[3) 4)]							

Figure 2-12. GTP header format. (Source: IP for 3G. Reproduced by permission of © 2002 John Wiley & Sons Ltd.)

sent using UDP over IP using the IP address of the tunnel endpoint – e.g. the GGSN for traffic sent from the SGGN to an external network. The most important header field is the tunnel id (identifier) which identifies the GTP packets as belonging to a particular PDP context of a specific user (and therefore can be given the appropriate QoS). The tunnel id is formed from a combination of the IMSI and NSAPI – the IMSI uniquely identifying a terminal and the NSAPI being a number from 0 to 15 that identifies the PDP context or the secondary PDP context within a primary PDP context.

In the UMTS core network IP layer 3 routing is, typically, supported by ATM switching networks. It is the operator's choice whether to implement QoS at the IP or ATM level but if the IP layer is used then the IETF differentiated services scheme is specified by 3GPP as the QoS mechanism. In all cases interoperability between operators is based on the use of Service Level Agreements that are an integral part of the definition of DiffServ.

If DiffServ is being used to provide QoS in the core network then a mapping is needed at the RNC between UMTS bearer QoS parameters and DiffServ code points and a similar mapping is needed at the GGSN for incoming packets. The SGSN originally downloads the user subscription data from the HLR and passes the allocation/retention priority, firstly to the RNC with the Radio Access Bearer request and then to the GGSN with the PDP context activation message. The RNC and GGSN then use the allocation/retention priority and the UMTS class to map to DiffServ classes as shown in Figure 2.13. In Diffserv there is no delay bound and such a network would rely on proper provisioning to deliver sufficiently low delays for conversational services.

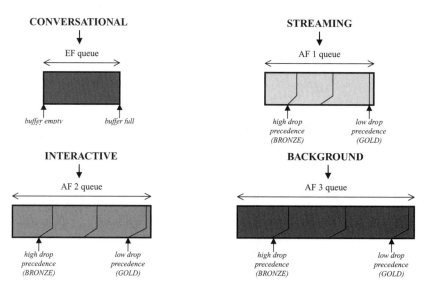

Figure 2-13. Mapping of UMTS QoS Classes to DiffServ queues. (*Source: IP for 3G.* Reproduced by permission of © 2002 John Wiley & Sons Ltd.)

UMTS decouples the terminal packet data protocol from the network transport – through the use of tunnelling. As a consequence it can transport IPv4 or v6 packets without modification. As was explained above – in practice no operators are using such a QoS scheme for mobile data at the present time. VoIP and real-time services are not offered over the PS network[4] – all the traffic is background. Video is normally sent over the CS network. In any case the congestion that is now seen in 3G networks is in either the air interface or from the Node Bs to the RNCs – these are often connected by one to three E1 (2Mbit/s) leased lines. It is said the backhaul costs are 30% or so of the annual running costs of a mobile network (OPEX). If there is congestion beyond the air interface this is where it will be.

2.5.7 Signalling in the UMTS Core Network

The signalling between the mobile, SGSN and GGSN to the HLR, authentication centre, EIR and also the SMS message centre all consist of SS7 signalling (Signalling System Number 7) messages (see Figure 2.5 again). SS7 is, in some ways, like an IP network (but it is not IP at all and developed totally independently!): it is packet based, has reliable transport protocols and its own addressing scheme. SS7 was originally used on the PSTN when it became digital and carries all the signalling messages between exchanges needed to set up a call (address complete, ringing, connecting . . . being example messages). The SS7 variant used in the PSTN is called ISUP and this has been extended for use in mobile networks with an extended message set called MAP (Mobile Application Part). The SGSN/HLR and so forth all have SS7 addresses and use MAP to exchange signalling messages. From the SGSN the G_r interface connects to a (logically separate) SS7 network over a 2 Mbit/s time division multiplex link (the normal circuit switched connection you would find in the PSTN say – i.e. not IP and totally separated from the data path transmission mechanism).

SS7 is not the only signalling protocol used in the UMTS core network. The setting up, modifying and tearing down of GTP tunnels is performed by a signalling protocol called GTP-C (whilst the transport of user data is performed by GTP-U which we have just described). GTP-C runs between the SGSN and GGSN and also caries the messages to set and delete PDP contexts. GTP-C uses the same header as GTP-U but is a reliable protocol in that the sequence numbers are used to keep track of lost messages and these are re-sent. An example GTP-C message is ECHO – this can be sent to another GSN that must reply with an ECHO RESPONSE message that includes the time since the last re-boot. Readers needing more details of GTP-C messages are referred to the TS 29.060 where the nitty gritty detail awaits! There is no SS7 signalling link from the SGSN to the GGSN.

Note that GTP-C does not run over the I_u interface between the SGSN and the RNC – since RNCs have no part in PDP context activation etc. – the GTP tunnels

---

[4] Skype – offered as part of a package on Internet services by "3" is carried over a 3G circuit.

from RNCs to SGSNs are set up by part of another protocol RANAP – this is covered in the next section.

2.6 UMTS Radio Access Network (UTRAN) and Air Interface

In this section we will look more closely at the interfaces of the UTRAN, the signalling and transport protocols used to convey bits from the mobile to the RNC and also look at the underlying ATM switching cloud. This is to illustrate just how un-IP friendly the UTRAN is. Firstly, let's look at the switching and timing requirements for soft handover in greater detail – as soft handover has been a double-edged sword for UMTS. Secondly, to bring out just how complicated the RAN is – with functions like Radio Resource Management requiring intensive, real-time, processing in a number of distributed elements Finally, the layer 2/layer 3 interface is very much integrated and tailored for W-CDMA (with its power control and particular radio characteristics). For all these reasons we will wade through: the air interface (U_u); the UTRAN to core (I_u) interface and, finally, tackle the ATM transport.

It is important to note that the standards concentrate on the interfaces because that is where equipment from different manufacturers needs to inter-operate; for example an Ericsson terminal needs to talk to any make of base-station correctly. Within a base-station, however, the operation of soft handover and the QoS scheduling is completely propriety and the standards do not specify these.

2.6.1 The W-CDMA Air Interface and the U_u Interface

CDMA stands for code division multiple access – meaning that many users share a single block of spectrum by means of different code sequences that they multiply (spread) their data with to increase the bit rate prior to transmission. For example, if I have a 38.4 kbit/s data stream and spread it with a chip rate of 3.84 Mchip/s (the spread code bits are called chips) then I have, somewhat obviously, a spreading factor of 100. Now the clever thing about CDMA is that if I multiply the spread signal by the same spreading code again I recover the original bit stream. Moreover, if there are other users with different spreading codes then the result of multiplying their transmission with my spreading code is simply noise – providing the codes are carefully chosen. This process is sometimes likened to an international party where you hear somebody over the far side of the room talking your language and lock on to that conversation above a general background noise created by other conversations. You might ask whether it would not be better just to divide the spectrum up and give each user his own part of the spectrum (frequency division multiple access) or time slot (time division multiple access). Many simulations have shown that CDMA can support more users at a given QoS and user bit rate (Chapter 4 provides an insight into this).

In UMTS there are two modes of operation FDD and TDD. As was already explained the FDD (Frequency Division Duplex) mode uses two blocks of spectrum for the up and down links and separates users solely on the basis of CDMA codes.

The TDD (the time division duplex) mode uses a single block of spectrum but only transmits on the up link or the down link at one time, hence the time division duplex, and uses a mixture of codes and time slots to separate users. We will concentrate on the FDD mode here because the TDD spectrum has still yet to be exploited by the mobile operators.

In a CDMA system neighbouring cells use the same frequency but avoid direct interference by means of the use of scrambling codes.[5] From the base-station to the terminal the spreading code is made up of two parts – the scrambling code that is different for each cell (or sector) and the channelisation code that separates users within the cell. Transmissions from users and base-stations in neighbouring cells are always seen as noise because the spreading code (=channelisation code × scrambling code) is unique for each base-station to user transmission within the entire system. This is how CDMA is able to use the same frequency in every cell.

In UMTS the CDMA air interface makes up the physical layer and part of the MAC layer of the U_u interface between the terminal and the base-station. The air interface (U_u interface) protocols are shown in Figure 2.14. The PDCP (Packet Data Convergence Protocol) provides header compression for PDP packets – as already described. The BMC (Broadcast/Multicast Control) layer provides cell broadcast facilities.

The RLC (Radio Link Control) layer is responsible for setting up and tearing down RLC connections – each one represents a different radio bearer (meaning there is one radio bearer per PDP context or circuit). The RLC layer segments and reassembles data packets as well as providing backward error correction. A 1500 byte IP packet would be segmented into 27 RLC PDUs with a 2 byte header added to each (the MAC layer would add another 3 bytes from MAC PDUs). The level of backward error correction can be one of several modes:

- Transparent – higher layer packets are not provided with error recovery and higher layer packets may be lost or duplicated.

- Unacknowledged – this mode detects errored packets but simply deletes them. It also avoids duplicating packets.

- Acknowledged – Error free delivery of packets is guaranteed by an ARQ (automatic repeat request) backward error recovery scheme. Also duplicate detection ensures that only one copy of each packet is transmitted.

The RLC is also responsible for ciphering and can perform flow control – i.e. the receiving end can request the transmitting end to slow down transmission to

---

[5] There are different W-CDMA carriers each of which occupies 5 MHz but one carrier can be used on adjacent base-stations or carriers.

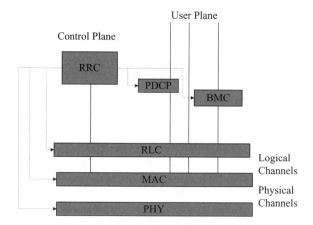

Figure 2-14. Radio Interface protocols – control and user plane. (*Source: IP for 3G. Reproduced by permission of © 2002 John Wiley & Sons Ltd.*)

prevent, for example, buffer overflow. For data the RLC terminates at the RNC – so RLC frames are carried to the Node B over the radio interface and MAC layers and thence on to the RNC on AAL2/ATM switched circuits (see below).

The MAC layer is responsible for mapping logical channels (including data flows) into the transport channels provided by the physical layer. Logical channels in UMTS (FDD mode) include:

- Common control channel (CCCH) – uplink.

- Broadcast control channel (BCCH) – downlink.

- Paging control channel (PCCH) – downlink.

- Dedicated Control Channel (DCCH) – dedicated (to a single terminal) transport channel (up and down link).

The MAC layer is also responsible for multiplexing/demultiplexing flows from the user onto transport channels (that are similar, but fewer in number to the logical channels – e.g. a BCH (Broadcast Channel carries the contents of the BCCH). The MAC also handles priority handling of flows from one user – i.e. allowing flows with higher priority QoS to have higher priority access to physical channels.

The physical layer is responsible for transmission of data blocks: multiplexing of different transport channels (e.g. the P-CCPCH – Primary Common Control Physical CHannel carries the BCH), forward error correction (error coding) and error detection, spreading (with the CDMA code) and RF modulation.

Much, much more detail on the UMTS CDMA physical layer can be found in Ref. 10. Logical and transport channels are covered further in Chapter 4.

2.6.2 UTRAN Mobility Management

2.6.2.1 Soft Handover

The requirement to support soft handover in UMTS arises from the handover of mobiles between base-stations. The boundary between the cells is not a clean dividing line – by which I mean that when you are anywhere near the boundary the ratio of received power from the two base-stations fluctuates quite considerably over even a metre or so (at 2 GHz the wavelength is 15 cm). If you set the handover threshold to be when the new base-station received strength exceeded the old base-station strength then the handover would be pinging back and forth all the time. Now each handover "costs" in terms of signalling messages and network processing time so, to avoid all the ping ponging, the handover threshold is given hysteresis: once a handover has occurred the relative signal strengths must change by 6dB, say, before another change is made. This is fine for TDMA systems (like GSM) – where the interference is felt in distant cells that are re-using that frequency. However, in CDMA systems, having mobiles operating at 6dB over their minimum power causes a lot of interference – in CDMA systems all mobiles interfere with each other and controlling and minimising transmit power is the key to increasing capacity. It has been estimated that using the TDMA hysteresis scheme for handover would reduce the efficiency of a UMTS system by 50%. The solution is something called soft handover. In soft handover (Figure 2.15) the mobile receives transmissions from several base-stations simultaneously. As the power – and hence the error rate – from each fluctuates the mobile takes the received data from each base-station and combines them to get a reliable answer.

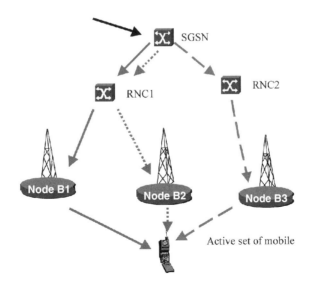

Figure 2-15. CDMA soft handover. (Source: IP for 3G. Reproduced by permission of © 2002 John Wiley & Sons Ltd.)

It has been estimated that UMTS mobiles are in soft handover 50% of the time and will be connected to several base-stations simultaneously.

For soft handover to work it requires that the frames from different base-stations arrive at the mobile within about 20 milliseconds or so of each other. Thus from the network split point they must be transported to the transmitters with very tight control of delay and jitter. ATM gives this functionality as the transport technology for the UTRAN.

2.6.2.2 Handover Types

As we have seen above in a CDMA system users are connected to a number of cells – called the active set – and cells are added and dropped from the active set on the basis of measurements made by the terminals and reported back to the network. If the cells are served by the same base-station (Node B) then the mechanism of adding/dropping cells from the active set is proprietary – i.e. the standards do not specify how it shall be accomplished – this is softer handover.

Now imagine that a user needs to connect to cells on a different RNC (Figure 2.16). The original RNC – called the serving RNC – connects to the new RNC – called the Drift RNC – via the I_{ur} interface. This interface has no counterpart in GPRS or GSM and allows the UTRAN to deal with all handovers independently of the core (this is required in CDMA because of the tight timing constraints for soft handover and the need to add and delete cells from the active sets rapidly). At some point the UTRAN decides that it should move the SGSN to RNC connection from the Serving to the Drift RNC – a process called SRNS (Serving Radio Network System) relocation. This is essentially a UTRAN function and the result of the procedure is that the SGSN routes the packets to the new RNC.

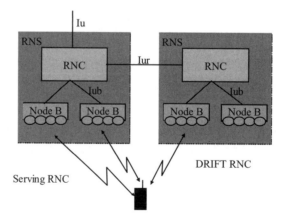

Figure 2-16. Change of RNC – intra-SGSN SRNS relocation. (Source: IP for 3G. Reproduced by permission of © 2002 John Wiley & Sons Ltd.)

If a user moves within the coverage area of a base-station (and RNC) served by a different SGSN – then this requires the highest level of mobility management – the Inter-SGSN/MSC SRNS relocation (we will concentrate on the packet case). Figure 2.17 shows the situation both before and after SRNS relocation. This is a complex procedure – involving the mobile, 2 RNCs, 2 SGSNs and a GGSN! Readers wishing to see the message flows and sequence of events can look it up in 3GPP standard TS 23.060 (downloadable from

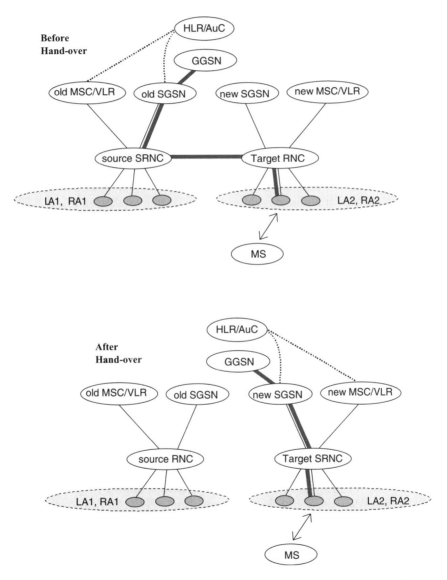

Figure 2-17.   Change of RNC – inter-SGSN SRNS relocation. (Source: *IP for 3G.* Reproduced by permission of © 2002 John Wiley & Sons Ltd.)

www.3gpp.org). One noteworthy point about this procedure in R99 is that it incurs long delays in handing over packet connections. This has been fixed in subsequent releases of the standard. For circuit switched traffic moving to a new MSC there is, of course, no gap in service.

2.6.3 UTRAN Transport

Within the UTRAN all user data transmission takes place over an ATM switched network. Now ATM was originally conceived for fixed networks as a replacement (or evolution) of the PSTN/ISDN to allow packet and circuit data with varying traffic characteristics (constant or variable bit rate) to be multiplexed onto a single connection with guaranteed QoS performance. It was designed to run over optical links with characteristically low error rates (Bit error rate less than 10^{-9}) and so had very light error correction. ATM controls delay and jitter of traffic by carrying all the data inside 53 byte cells (fixed length packets). Because all the cells are the same size very efficient cell switches could be made that could control the jitter and delay of the cells being switched. Typically virtual connections would be set up through a mesh of ATM switches (an ATM cloud) by rather slow signalling – hence the term permanent virtual circuits (PVCs) – and voice/IP/video packets or frames or streams would be adapted, using different ATM adaptation layers (AALs), to be segmented and reassembled at the ends of the PVC. The AALs differ in how they segment/reassemble higher layer packets and in the error checking and recovery mechanisms that they provide.

In UMTS AAL2 is used for all circuit and packet data within the RAN whereas AAL5 is used for signalling within the UTRAN and for transmitting the packet data across the $I_u(PS)$ interface to the SGSN. AAL5 has very little functionality: other than segmenting and reassembling packets into ATM cells; it provides only a basic error check.

Why AAL2 is used for transport within the UTRAN can be traced back to the particular requirements of CDMA operation and also the desire to support multimedia traffic. 2G's 64 kbit/s circuit switching technology is clearly inefficient for bursty Internet type traffic – what is needed is a packet switching technology that multiplexes together many users and uses less underlying (and very expensive) bandwidth in the access network (i.e. the leased lines that typically connect the UTRAN). The requirement from CDMA, as readers will remember from earlier, was that in order to use spectrum efficiently a CDMA system had to exercise very tight power control. This means that it has to support soft handover – with terminals connected to several base-stations simultaneously. More importantly the combination of signals takes place at layers 1 and 2 and the signals need to arrive within 10 msec or so of each other for correct combination. RNCs need to be able to setup and tear down connections to Node Bs and other RNCs to add and remove cells from the mobiles active set (of cells it is in contact with). This set up needs to be accomplished rapidly – within 100 msec or so. Finally, the desire to transmit voice, pretty essential for any mobile network, meant that only small packets could be used in the access network. The total delay for a voice call should not exceed about 100 ms. At 12 kbit/s, allowing 20 msec for forming a packet – the packetisation delay – gives a packet size of 240 bits or 30 bytes. An ATM cell carries 48 bytes (compared to an IP packet typically 1500 bytes – which is one of the reasons why Voice over IP – VoIP – is inefficient). The AMR (adaptive multi rate) speech coders used in UMTS networks typically produce a variable bit rate – for example they don't code silence – so a constant packetisation delay requires variable length packets. When the coder is only producing 4bit/s I need smaller packets than when it is producing 14 kbit/s – for the same packetisation delay.

From these requirements, along with the need to provide QoS, a new ATM adaptation layer – AAL 2 – and its associated switching and signalling procedures has been developed for the UMTS radio access network. AAL2 allows variable length packets and multiplexes several connections onto a single ATM Virtual Connection. AAL2/ATM is used to carry all the user data – packet and voice – over the UTRAN – the packet data only being converted back to AAL5 at the RNC. The need to multiplex many different, variable rate, traffic sources onto a single ATM VC was the reason for choosing AAL2. Essentially the key part for UMTS was the development of signalling and switching of AAL2 circuits. A very comprehensive review of AAL2 for UMTS is given in Ref. 11.

2.6.4 UTRAN QoS

When a user wishes to make a call or send packets then control signalling for this first passes to the MSC/SGSN. In the packet case there must be a PDP context active – so that the terminal has an IP address and the GTP tunnel id is allocated. However the PDP context does not install any state within the UTRAN and so each time packet transfer or a connection takes place radio and UTRAN bearers (see Figure 2.7) must be allocated. The SGSN/MSC signals the UTRAN with the required QoS attributes and these may either be granted, re-negotiated or declined. The control protocol (RANAP) used between the SGSN and RNC is described in the next section.

QoS provision within the UTRAN is quite complicated and can be broken down into air interface and RAN parts. In a well designed network the air interface will be the major bottleneck – where most congestion and QoS violation will take place. A Radio Resource Management (RRM) function that is distributed between the terminal, base-station and the RNC controls QoS over the radio link. RRM consists of algorithms and procedures for the following:

- Admission control;
- Power control;
- Code management;

- Packet scheduling;

- Handover.

In a CDMA system the common resource consumed as more users are admitted to the system is interference to other users. As users transmit progressively more power then the higher the bit rate or lower the error rate that they can achieve – but at the expense of causing higher interference to all other users within the cell and in neighbouring cells (in W-CDMA it is mostly same cell). The RRM system must ensure that, even after admitting a new connection or packet stream, the overall interference will be low enough to satisfy both the new and existing QoS requirements. Typically UMTS will run at about 50% of its maximum capacity to ensure stability.

CDMA codes are managed by the RNC – in the downlink channelisation codes must be allocated to each user (so that the spreading code is the cell scrambling code x channelisation code). The codes for each user must be orthogonal (i.e. have no correlation when multiplied together). In the uplink the scrambling codes are used to separate users and channelisation codes to separate data and control channels from a single user. All these codes are allocated/de-allocated and managed by the RNC.

Packet scheduling takes place each time a user or a base-station has packets to transmit (other than very small amount of data that can be transmitted on the random access channel). A request is made to the RNC and only admitted if resources are available. These requests can be queued – and how they are treated depends on the allocation/retention parameter and used by the RNC when deciding how to allocate resources when faced with new QoS resource requests. The idea here is that operators can offer different priority levels for resource allocation – even for users who have requested the same QoS. As an example I might be a gold user (business class) – for which I pay a large subscription and which results in an appropriate allocation and retention policy entry in the HLR against my name. When I want to transmit some packets a request arrives at the RNC and is treated with the respect it deserves! – i.e. it goes to the head of the request queue and even if no resources are available it causes lower priority users – such as bronze (economy class) – to have their resources reduced. In real terms, I get to make my important VoIP call to the chairman of selectors at Lords and some impoverished student loses half the bandwidth of his video download of Coldplay – as it should be!

The allocation/retention parameter is complex and contains information on: priority, pre-emption capability, pre-emption vulnerability and whether the request can be queued. The resource scheduling algorithm, located in the RNC, is not specified by 3GPP and is vendor implementation specific. The final reply from the RNC is either success/failure or failure due to timing out in a request queue.

One proposal for a practical, flexible way of providing QoS to users for Internet services is the concept of services classes so this is a sub-division of the background traffic class. Let us assume we have three classes – gold, silver and bronze. Each class offers a certain group behaviour to users of that class. An example would be:

- Gold users always get requested bandwidth – regardless of interference, congestion or radio degradation (unless they themselves are already using all the available resources).

- Silver users have an elastic bandwidth but the service is better than that experienced by bronze users who get the equivalent of a best effort service.

- Bronze users get "bog all".[6]

Of course different operators will operate different QoS schemes and the standards do not specify exactly how QoS is achieved or implemented in UMTS – just how it is signalled across the interfaces. In practice today QoS is only used to allocate resources to CS calls and then share what is left between users of mobile broadband which has only a background class associated with it. Some MNOs separate data and voice by using one carrier for voice and others for data.

The radio bearer is also responsible for mapping the appropriate class to the correct error correction mechanism. Conversational services tolerate little delay and need forward error correction (redundancy added to the transmitted frame) as opposed to backward error correction (re-transmission of corrupted/lost frames) which causes delay.

Handover in UMTS is handled by the RRM. As we have seen earlier this includes softer handover (when handled within a base-station) and soft handover – where the terminal has an active set of base-stations that it is in contact with. The RRM decides when base-stations are added or deleted from the active set – sometimes this is suggested by the terminal but the network always has control of this process.

2.6.5 UTRAN Signalling

The signalling across the Iu interface (from SGSN to RNC) is provided by RANAP – Radio Access Network Application Part. RANAP is responsible for:

- Radio Access Bearer set up, modification and release;

- Control of the UTRAN security mode;

- Management of RNC relocation procedures;

- Exchanging user information between the RNC and the Core Network;

---

[6] Contribution from a top researcher in BT's Network Lab!

- Transport of mobility management and communication control information between the core network and the mobile. (The so called Non-Access Stratum(NAS) information – such as PDP context management and IP address– that does not concern the UTRAN);

- Set up of GTP tunnels between the SGSN and the RNC.

From the RNC to the terminal the Radio Resource Controller (RRC) (see Figure 2.14) sets up a signalling connection from the user equipment (UE) to the RNC. This covers the assignment, re-configuration and release of radio resources. The RRC also handles handover, cell re-selection, paging updates and notifications.

2.7 CDMA2000 Packet Core Network

cdma2000 is another member of the IMT-2000 family of 3G standards and has a North American origin. Originally there was cdmaOne – which utilised the IS-95A CDMA air interface and an ANSI-41 network (ANSI-41 is another, GSM-like, network of base-stations, base-station controllers and specifies the protocol used to signal between them). CdmaOne was launched in Hong Kong in 1995 and is now widely used for voice and low bit rate data in the Far East and North America. The first upgrade to cdmaOne was a new air interface – IS-95B with data rates up to 64kbit/s being offered by some operators. Basic cdma2000 1X is designed to be backwards compatible with cdmaOne, offers twice the voice capacity (in the same spectrum) and data rates up to 144kbit/s. Beyond 1X cdma2000 there are a number of revisions (0/A/B) that have been standardised and will be described in Chapter 4. EVDO (Evolution Data Only) offers higher downlink data speeds and Rev.A even higher data speeds – all the details are in Chapter 4 – but I am telling you this to make sense of the statistics below. According to the CDMA development group (http://www.cdg.org/) there are:

- 250 commercial operators;
- 99 countries;
- 246 commercial 1X networks;
- 28 1X networks in deployment;
- 82 commercial 1xEVDO Rel. 0 networks;
- 53 1xEVDO Rel. 0 networks in deployment;
- 24 commercial 1xEVDO Rev. A networks;
- 30 1xEVDO Rev. A networks in deployment;
- 400 400 000 CDMA2000 subscribers (3Q 2007);
- 83 000 000 CDMA2000 1xEVDO subscribers (3Q 2007).

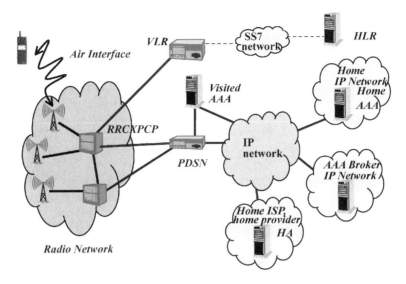

Figure 2-18. PCN – Packet Core Network architecture. (Source: IP for 3G. Reproduced by permission of © 2002 John Wiley & Sons Ltd.)

The packet core network (PCN) of cdma2000 is shown in Figure 2.18 – the CS part (not shown) is simply corrected to an MSC and HLR as in UMTS – because the operation of this is very different to UMTS.

Much like UMTS this includes a radio access network (and all the concomitant issues of soft handover) as well as a PDSN (Packet Data Service Node) – roughly equivalent to a SGSN. The major difference between the PCN and UMTS is in the way that mobility management is handled. In UMTS, as we have seen, this is handled in the HLR and uses SS7 signalling – the PCN, however, is based on Mobile IP (MIP) – an Internet mobility concept.

MIP has been developed in the IETF because operating systems and applications are not always tolerant to a change of IP address. Although XP, for example, allows a change of interface and access technology (e.g. Ethernet to WLAN) the time needed for a handover is often 30 sec or so. Some applications (mostly client-server) such as web browsing and email survive the change. Others, such as ftp, fail (smart ftp can be used) and real-time applications (VoIP and video) always freeze.

In MIP the home agent (HA) – which receives all packets sent to a users home address. When the user is roaming on a foreign network it is allocated a care-of address by a foreign agent (FA) (a software process that sends out advertisements and responds to requests for addresses). The roaming terminal then tells its home agent what its care-of address is and when packets arrive, addressed to the home address, the HA tunnels them – using IP in IP tunnelling – to the FA. The FA decapsulates the packets and sends them on the mobile user using a layer 2 address (remember the FA and mobile are on the same subnet). Packets from the mobile to correspondent host (CH) can be sent directly – since they have a destination address that routers can use directly.[7] This is MIP – and the great beauty of it is that it does not require any change to existing routers, routing protocols or existing IP stacks that are not mobile-aware. The best analogy is the postal one already described in tunnelling – I send my letters (packets) to your home address and your mother (home agent) puts them in a new envelope (encapsulates them) addressed to your college address (care-of address). When the letters arrive you (acting as a foreign agent) tear open the envelope and find the original letter inside. You keep in touch with your mum by separate letters (signalling) that include notification of your new address (registration) and your mum knows it is you because she recognises your handwriting (authentication!). All these functions are present in MIP.

The PDSN acts as a foreign agent – providing care-of addresses to mobiles and decapsulating IP packets tunnelled from the HA. The link from the PDSN to the mobile is made using PPP (point to point protocol) – PPP is more familiar from dial-up networks where it provides encapsulation and error protection from computers to the NAS (network access server). The radio interface between the RN and PDSN has two channels – one for data and the other for signalling. The signalling comprises standard MIP messages (registration request, registration reply) plus two additions – registration update and registration acknowledge. Terminals, therefore, have to run a cdma2000 special IP stack – containing the non-standard MIP code. MIP does nothing to provide mobility in the radio network and, since this must support soft handover, this is not based on IP but rather ATM. Basically the additions to the standard MIP framework are:

- The use of an AAA (Authorisation, Authentication and Accounting) server;

- The packet data related RADIUS attributes;

- Use of the PDSN node as the FA.

The AAA (Authorisation, Authentication and Accounting) server is a standard IP server carrying details about subscribers and is used to authenticate users and check that they are able to use the requested service; in this respect it is performing a similar role to the HLR in UMTS. It also stores and forwards accounting information – usage data records generated by the PDSN.

The PDSN acts as both a mobility anchor and operates as a RADIUS client in forwarding authentication details towards the appropriate AAA server. The RADIUS protocol – Remote Authorisation Dial In User Service – has been extended to carry additional attributes specific to cdma2000 – typically session status, differentiated service class options and accounting attributes.

---

[7] Many real-life systems don't allow this because the "from" IP address does not belong to the domain from which the node is sending – rather it is the home IP address of the node. Firewalls often block packets sent with such IP addresses because IP address spoofing is a common technique for all sorts of security attacks on networks. The solution used by MIP is reverse tunnelling – where the packets from the mobile node are tunnelled back to the home agent and then sent to the CH.

A typical MIP registration in cdma2000 would comprise:

- A mobile roams to a new network and sends a request for service to the base-station, the base-station checks for existing links and then forwards the request to the PDSN.

- The mobile negotiates with the PDSN using PPP and sends its id and security data to the PDSN.

- The PDSN uses the id to locate the home AAA and places the security data into a RADIUS request.

- If the home AAA authenticates the user the PPP link is finally established.

- The mobile node then solicits for a FA – the PDSN responds with an advertisement showing available foreign agents.

- The mobile creates a MIP request and obtains a FA care-of address.

- The mobile forwards its care-of address – probably securely – to the HA which creates a tunnel for any incoming packets.

If the mobile moves from the area of one base-station controller to another, and they are both connected to the same PDSN, then the PDSN is able to associate the mobile with the old PPP state – this is an intra-PDSN handover. If the mobile moves into the area covered by a new PDSN then it has effectively roamed to a different foreign network and must establish a new PPP connection, obtain a new care-of address and register again with its HA.

2.8 TD-SCDMA

As was mentioned in the introduction the Chinese have developed a Time Division-Synchronous Code Division Multiple Access system. This is a TDD system – using the same frequency block for up and downlinks which brings with it a number of advantages. Firstly, operation in small spectral bands (as low as 1.6 MHz) is possible. Secondly, the up/down split is flexible – meaning that if much more traffic is downloading web pages and files then more of the capacity can be allocated to that. Finally, TDD systems are better suited to smart antenna techniques we will meet in Chapter 5 on WiMAX. TD-SCDMA also uses a combination of TDMA and CDMA to separate individual users. Uplink transmissions are synchronous (the "S" in TD-SCDMA) – meaning each terminal is tightly time controlled by ranging – which (as will be explained in Chapter 4) means the uplink codes are orthogonal and interfere less (in cdma2000 and UMTS they are not synchronised).

It has been said (e.g. http://en.wikipedia.org/wiki/TD-SCDMA) that the main reason behind the initiative is to avoid the heavy patent costs associated with other 3G technologies.

2.9 Conclusion

3G was conceived in the late 1980s to meet the "martini" vision – communications "anytime, anyplace, anywhere". However, the 3G concept has undergone radical changes since then – the satellite component, the B-ISDN and the fixed mobile convergence ideas have all been removed. The industry – operators and manufacturers – have also chosen an evolutionary route for 3G core networks, leveraging their extensive investment and research in 2G networks. The reasons for this include the runaway success of 2G voice networks – to the point where costs are now rivalling fixed-line operations for voice traffic. GSM and other 2G technologies were the products of a tight standardisation process – where complete, integrated, solutions were specified by means of interfaces. This process has also been adopted for 3G-standardisation.

In sharp contrast the air interface(s) chosen are revolutionary and the decision appeared to be, to many neutral observers, a political fudge that allowed five different standards to be called 3G. The nature of CDMA, requiring soft handover support, and the need to multiplex variable rate voice and multimedia traffic has also led to the adoption of the transport technology of the 1980s – ATM – for the radio access network.

Whilst 3G has been moving through the standardisation process a second great revolution, after 2G mobile, has been unfolding – the Internet. Nowadays nearly all data transmissions, and an increasing number of voice calls, are encapsulated within IP packets before leaving the end terminal. IP services have grown rapidly, with email, browsing and countless business applications built on IP technology.

Of course 3G has been very slow to take off and where it has been deployed the bulk of the traffic has been voice and messaging. In Chapter 4 we will look in depth at the service side of 3G since launch as well as seeing how the decisions during standardisation and its general non IP approach have held it back during a time when fixed broadband has exploded and the Internet has become all-pervasive. We will then see how updates to the standards are tackling some of these issues. However, first we must look at another technology that is all-IP, offers up to 100 Mbit/s and can be had for free on most high streets – Wireless LANs.

References

[1] Huber, J. et al., UMTS, the Mobile Multimedia Vision for IMT-2000: A Focuson Standardization, IEEE Personal Communications, September 2000, pp. 129–136.
[2] "Third generation Mobile Systems", A. Claptonand I. Groves, *BT Technology Journal*, Vol. 14, No. 3 July 1996, pp. 115–122.
[3] "Towards the personal communications environment: green paper in the field of mobile and personal communication in the EU" – CEC COM(94) 145 (27 April 1994).
[4] "Global Multimedia Mobility – A Standardization Framework", ETSI PAC16 (96) 16 Parts A/B June 1996.

[5]   "The Evolution of UMTS/HSPA Around the World", Alan Hadden, President, Global Mobile Suppliers Association, Informa Telecoms & Media – LTE World Summit, Berlin, 21–23 May 2008. see www.gsacom.comInforma

[6]   "Radio spectrum for mobile networks", T. Hewitt, *BT technology Journal*, Vol. 14, No. 3 July 1996, pp. 16–286.

[7]   Mike Walters – Vodafone – comments at the Informa Telecoms & Media -LTE World Summit, Berlin, 21–23 May 2008.

[8]   "The GSM System FOR Mobile Communications", M. Mouly and M. Pautet, Bay Foreign Language Books, 1992, ISBN-10: 2950719007.

[9]   3GPP standards 33.900 and 33.102 – available from www.3gpp.org

[10]   "Understanding UMTS Radio Network – Modelling, Planning and Automated Optimization", M. Nawrocki, M. Dohler and A. Aghvami, John Wiley & Sons Ltd, 2006, ISBN-13 978-0-0470-01567-4 (HB).

[11]   "Applying ATM/AAL2 as a switching technology in third generation mobile access networks", *IEEE Comms. Mag.*, June 1999, pp. 112–122, G. Eneroth *et al.*

More to Explore

3G standards development

"The development of mobile is critically dependent on standards", F. Harrison and K. Holley, *BT Technology Journal*, Vol. 19, No. 1 Jan. 2001, pp. 32–37.

"The Complete Solution for Third-Generation Wireless communications: Two modes on Air, One Winning Strategy", M. Haardt and W. Mohr, IEEE Personal Comms, Dec. 2000, pp. 18–24.

"Third Generation Mobile Communications Systems", Eds R. Prasad, W. Mohr and W. Konhauser, Artech House 2000.

"Battle of the standards", Lynnette Luna, *Telephony* 19 Feb. 2001, pp. 62–70.

"Evolving from cdmaOne to third generation systems", G. Larsson, *Ericsson Review*, 2/2000, pp. 58–67.

"Harmonization of Global Third-Generation Mobile Systems", M. Zeng *et al.*, IEEE Personal Comms, Dec. 2000, pp. 94–104.

"Global Multimedia Mobility – A Standardization Framework", ETSI PAC16 (96) 16 Parts A/B June 1996.

Manufacturer information

www.ericsson.com – Ericsson Review available on line

Nokia Siemens Networks – excellent 3G tutorial (out of date but 6Mbyte and 126 slides): http://www.nmscommunications.com/DevPlatforms/OpenAccess/Technologies/3G324MandIPVideo/3GTutorial.htm

Mobile standards bodies and organisations:

3GIP – www.3gip.org– IP pressure group for UMTS

3GPP2 – www.3gpp2.org –cdma2000 standards

3GPP – www.3gpp.org – UMTS standards and technical specifications

UMTS Forum www.umts-forum.org – Information on UMTS from suppliers and operators

ITU – www.itu.int and www.itu.int/imt for imt-2000

IETF – www.ietf.org – Internet drafts and RFCs

OMA – http://www.openmobilealliance.org/

IEEE – http://www.ieee.org/portal/site

TD-SCDMA forum http://www.tdscdma-forum.org/EN/index.asp

GPRS

"GPRS Support Nodes" – L. Ekeroth and Per-Martin Hedstrom, *Ericsson Review*, No. 3, 2000.

"GPRS: Architecture, interfaces and deployment", Y. Lin, H. Rao and I. Chlamtac, *Wireless Comms and Mobile Computing*, 1, 2001, pp. 77–92.

"GPRS General packet radio service" – H. Granbohm and J. Wiklund, *Ericsson Review*, No. 2, 1999.

UMTS Standards

Available from www.3gpp.org

TS25.401 V3.70 – R99 UTRAN overall description

TS 23.060 v3.80 – R99 GPRS description (including GPRS for UMTS)

TS 23.413 v3.60 – R99 UMTS Iu interface RANAP signalling

TS 33.900 – A guide to 3rd generation security

TS 33.102 – 3G Security; Security Architecture

UMTS

"UMTS Network: Architecture, Mobility and Services", Heikki Kaaranen, *et al.* -John Wiley & Sons Ltd

"The UMTS Network and Radio Access Technology", J. Castro, John Wiley & Sons Ltd, 2001, ISBN 0 471 81375 3

BT Technology Journal – Vol. 19, No. 1 Jan. 2001 – Special issue on Future Mobile Networks

3G tutorial –http://www.umtsworld.com/technology/overview.htm or www. iec.org/tutorials/umts/topic01.html

UMTS QoS

"Supporting Ip QoS in the General Packet Radio Service", G. Priggouris *et al*, IEEE Network, Sept/Oct 2000, pp. 8–17.

Cdma2000

"The cdma2000 packet core network", *Ericsson Review* No. 2, 2001, Tim Murphy.

"Evolving from cdmaOne to third generation systems", *Ericsson Review* No. 2, 2000, Gwenn Larsson.

http://www.cdg.org/ – CDMA development group

Chapter 3: Wireless LANs

3.1 Introduction

Wireless LANs – which includes all things labelled WiFi – are everywhere today. If you buy a smart phone or PDA the chances are it has WiFi capability. If you've bought a laptop without WiFi in the last few years then you were done. And woe betide you if you haven't got a home WLAN network with all your family able to roam the furthest wilderness of the garden whilst booking tickets to Alaska. Even games consoles, like the Nintendo Wii, now demand a WiFi connection to keep up to date. And then there are all those hot spots – a while back a tongue in cheek survey suggested that most Americans thought a WiFi hot spot was what you got when you came home drunk smelling of perfume. Now you can't move in many US coffee shops for students, writers and professionals with forests of laptops. Wireless cities – with free or low cost wide area coverage – are springing up everywhere from Philadelphia to Amsterdam to Lancaster. Industry analysts are questioning whether WiFi will be the death of cellular mobile data (Ref. 1).

This chapter starts off gently looking at these and other applications in more detail. Along the way a number of issues are identified that are clearly causes for concern of users and operators of WLANs. In particular power consumption on WiFi-enabled handsets is awful (7–20 hours standby is typical!). Quality of Service is non-existent – when you are sitting in a hot spot making that important Voice over IP (VoIP) call and a group of students come in and start downloading the latest music and your call is ruined. Even more worrying – when your family is booking those airline tickets to Alaska what is to stop your neighbour downloading pornography over your DSL line via your WLAN?

In order to understand why WLANs have all these issues it is necessary to delve into the technicalities of how they work – which I have tried to do by

IP for 4G   David Wisely
© 2009 John Wiley & Sons, Ltd.

comparison with a fixed Ethernet. The chapter then surveys the types of WLAN available – with a handy guide to the "alphabet soup" of IEEE working groups and standards. This allows a discussion and comparison of proposals and initiatives to improve WLANs in areas like power consumption and security.

Finally, the chapter asks key questions about the commercial aspects of WLANs. Will WLAN coverage grow to the point where it is a serious rival to Cellular mobile data? Will everyone be using a WiFi-enabled phone in five years? Will Wireless cities actually make money for somebody? With the advent of higher capacity Cellular data through HSDPA and femtocells (see Chapter 4) these are really difficult questions to assess. And what will happen to WLANs if mobile WiMAX networks become common place – after all, WiMAX has been referred to as "WLAN on steroids" (Ref. 2) – won't it just replace WLAN?

3.2 Applications of WLANs

WLANs are not new – they appeared in the 1980s and were actively being researched by the author in the early 1990s (Ref. 3). Figure 3.1 shows a (now very cumbersome looking) integrated laptop with: Infra red short range communication (home made); WLAN (external card) and Cellular data (an analogue phone strapped on!). The big problem was that most equipment was proprietary and would not interwork – a Cisco card failed to recognise a Motorola Access Point, for example. In 1990 the Wireless Ethernet Compatibility Alliance (WECA) was established to test and ensure interoperability – endorsing the now familiar WiFi (Wireless Fidelity) logo to equipment that passed (WECA have now become the WiFi Alliance – Ref. 4). Now WiFi is totally synonymous with WLANs and the terms are (effectively) interchangeable.

Figure 3-1. The author with an early integrated WLAN/Cellular Mobile/Infrared Laptop (Ref. 3). (Source: Author. Reproduced by permission of © Dave Wisely, British Telecommunications plc.)

Figure 3-2.   Typical hot spot locations. (*Source:* BT. Reproduced by permission of © British Telecommunications plc.)

Hot spots (Figure 3.2) – public places with commercial WiFi available like cafes and hotels – are probably the most visible manifestation of WiFi interoperability. There were reported to be10 507 hot spots in the UK in the first half of 2006 – a 35% rise on two years earlier (Ref. 5). In Europe the total was 69 000 with 69% growth. Hot spots are mostly found in: cafes, hotels, pubs, airports, restaurants and train stations – Figure 3.3 shows the location of the 12 000

Figure 3-3.   BT WiFi hotspot locations. (*Source:* BT. Reproduced by permission of © British Telecommunications plc.)

reported BT Openzone capable hot spots in the UK and some affiliations (although only 2000 or so are actually owned by BT the rest are mostly owned by The Cloud – a hot spot network operator with several wireless ISPs running over the top). To access a hot spot you can either have a regular subscription (e.g. BT offers 500 min for £10/month – July 2008) or pay by the hour or day (20 pence per minute).

WiFi is big in the consumer space too – many DSL broadband ISPs are now offering a free wireless router for new subscriptions. A wireless router is able to share a single DSL connection to a number of wireless and fixed PCs – Figure 3.4 shows the BT Home Hub – along with a VoIP phone that can be used to make free or low cost calls over the Internet. Interesting the phone connects to the hub using DECT (the standard digital cordless phone technology). The BT home hub is part of a battle for "ownership" of the broadband customer whereby ISPs are trying to move beyond mere "commodity bit carriers" and offer higher value

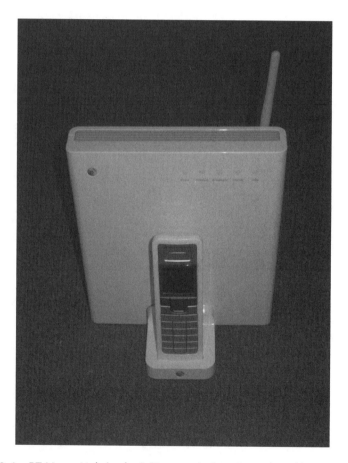

Figure 3-4.   BT Home Hub (author). (*Source:* Author. Reproduced by permission of © Dave Wisely, British Telecommunications plc.)

Figure 3-5. Access points advertised on British buses (Author's team). (*Source:* BT. Reproduced by permission of © British Telecommunications plc.)

services. BT has recently launched an IPTV service called "Vision" that connects to the BT home WLAN hub.

WLANs are also starting to appear on transport – with trains around the world slowly being upgraded for WLAN – Figure 3.5 shows that buses in the UK are starting to be equipped with them. In the UK more than a dozen train lines now have WLANs with, for example, WiFi having been launched on Heathrow Express early in 2007 (Ref. 6). WLAN have also been installed on some aircraft (Lufthansa) and are being developed by Boeing as the primary way of offering in-flight entertainment, voice calls and Internet access (Ref. 7).

Wireless cities are springing up everywhere. BT is currently installing WiFi in 12 (and later 20) UK cites including: Leeds, Cardiff, Westminster, Bognor Regis, Birmingham and Newcastle. More widely many US cities (Philadelphia, Seattle, San Francisco) and European (e.g. Amsterdam) and Far Eastern cities boast WiFi networks. Between them the geographic extent and degree of coverage within that area varies very considerably as does the cost to users. Some (particularly in the US) offer a basic free service – whereas others are available to subscribers of existing WLAN services (such as BT's Openzone) or on a day/hour basis. In London there are multiple competing systems with wholesale WLAN supplier The Cloud launching a mesh WiFi network across the square mile of the business district:

> "Customers will be able to take a taxi from one side of the square mile to the other and maintain a WiFi connection at all times", Owen Geddes, director of business development, told Total Telecom – Ref. 8.

Most of the wireless cities are meshed – meaning that only a small fraction of the access points are directly connected to the backhaul – the rest connect to gateway nodes using WLAN (sometimes on different channels or frequencies to avoid interference). Figure 3.6 shows the mesh concept as applied to BT Wireless Cities. In this case the phone boxes are the gateway nodes and various items of street furniture – such as lampposts – are the meshed nodes.

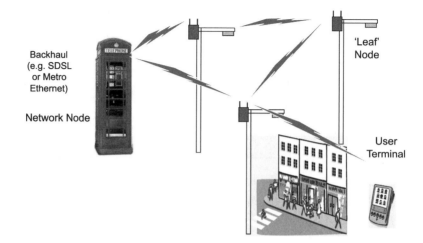

Network Node

Backhaul
(e.g. SDSL
or Metro
Ethernet)

'Leaf'
Node

User
Terminal

Figure 3-6.   Mesh WiFi used in Wireless Cities. (*Source:* BT. Reproduced by permission of © British Telecommunications plc.)

There are many other applications of WLAN – in particular VoIP (voice over IP) WiFi phones that can be bought at most computer stores or even supermarkets.

However – it is not all roses for WLANs and analysts reckon that very few players (outside of the equipment suppliers) have made any serious money from WLANs. For the next section the following quote from my 7-year-old son: "dad, do you want the bad news or the really bad news?" – seems very apt.

3.3 WLAN Compared to Cellular

Is it useful to compare a short range system like WLAN to a wide area system like 3G or GSM? Well I can offer two reasons for this. Firstly, that some people see WLANs growing in range and coverage (possibly with the help of WiMAX) to challenge cellular systems for traffic and customers and cellular becoming the technology of last resort when no WLANs are available. The proponents of this view suggest millions of access points (one in most homes and businesses) to cover a country like England. Secondly, WLANs are evolving to be more like cellular systems – with additions for things like: QoS, security and mobility. In the meantime cellular systems are becoming more like WLAN – with nano and pico base-stations and now femtocells (Chapter 4) being developed that cover WLAN-like areas inside the home/office.

Table 3.1 provides a (non exhaustive) comparison of the two. It is obvious that many of the differences stem from the respective standards bodies that gave rise to the two technologies. WLAN were specified and standardised by

	Cellular Mobile	WLANs
Coverage	More or less continuous Each base station can reach 100 m – 10 km	Limited to 10–50 m around an access point. Very high density of access points needed for continuous coverage
Speed	Low to middling (0.1–2 Mbit/ sbit/s – 2008 measurement in London with 3G and HSDPA)	Can be very high (5–25 Mbit/s) for an isolated access point but other users drastically reduce this as can small packets (required for VoIP)
Cost	Latest (July 2008) UK charges (see Chapter 7 for details) are £15/month for mobile broadband from Vodafone for 3G byte/month on 3G.	BT current prices are: 500 min for £10/month or 4000 for £25/month, 20 p/min PAYG or 24 hours for £10.
Terminals	Large range of handsets. PDAs and PC cards available	Mostly PC laptops, PDAs, a few cellular/wireless handsets available.
Battery Life	Excellent – up to 200 hours standby	Poor – 7–10 hours standby on dual WLAN/cellular handsets with WLAN turned on
Sign on	Strong security with SIM card – almost impossible to gain unauthorised access	Weak security – with name and password. Simple to gain access to many access points. (But note this is a deployment choice and could be improved)
Security of wireless traffic	Strong encryption of traffic and sign-on data	Weak or no encryption of traffic in hot spots but technology capable of high encryption levels.
Quality of service	Circuit switched traffic offers end-to-end guaranteed bandwidth. IP traffic prioritised with admission control and QoS classes (e.g. VoIP, video, WWW and background)	No QoS mechanisms and no admission control
Mobility	Full mobility with support for handovers and fast moving terminals	Basically nomadic support with handovers limited to enterprise systems with Access Points connected to a switch. Supports terminals at modest speeds (50 mph)
Interference	Managed within the system and controlled by the operator	Many causes outside the control of the access point operator (could be other users on same channel or other systems – e.g. video senders or external – e.g. microwave ovens). A serious problem for WLANs

Table 3.1.   A comparison of Cellular mobile and WLANs (as in most hot spots today).

	Cellular Mobile	WLANs
Spectrum	Exclusive use of licensed bands. Frequency use planned and monitored.	Licence-exempt spectrum – used by other systems and subject to strict power output rules. Frequency use random and unplanned. Limited number of channels available. (More at 5 GHz)
Services	Voice and SMS predominantly – some data services – WWW browsing and music download, limited TV	Mostly Internet and VPN (Virtual Private Network - connection to Enterprise Intranet). Some music downloads but limited voice.
Standards Body	3GPP and 3GPP2	IEEE

Table 3.1 (Continued)

the IEEE, as was noted in Chapter 2. Cellular mobile was mostly the work of 3GPP (and 3GPP2) these bodies tend to be dominated by mobile industry giants and produces specifications for complete solutions that encompass all aspects of the system. The IEEE, however, reflects a computer, IETF, beard and sandals type approach to standardisation. WLANs, as specified by the IEEE, are only ever part of the story – much more needs to be added to make them comparable to Cellular Mobile (e.g. billing, routing, security after the access point, mobility support, QoS). Table 3.1 gives a comparison of cellular mobile with a typical WLAN hot spot today.

All of these differences can be traced to the original design intentions of WLAN – as a wireless version of Ethernet – and the design philosophy of the IEEE. The next section will explore these with a (limited – you will be relieved to hear) description of how WLANs work at a nuts and bolts level. Then the chapter will continue with a look at how various bodies are adding to basic WLAN functionality as they evolve to be more like cellular systems.

3.4 How WLANs Work

This is a good time to recall the OSI/IP layer model (Figure 3.7). The good news is that WLANs are only specified at the physical layer and part of the data link layer. The rest – IP routing, session control etc. is outside the scope of WLANs and sits above them. This in itself is actually one of the reasons why security and QoS on WLANs is hard – since these are clearly needed end to end and so an IP or higher layer solution needs to interface to the WLAN capabilities (if any). We will tackle WLANs in the order MAC layer and then PHY layer – mainly because all the problems with WLANs mostly relate to the MAC layer. The PHY layer is really about the transmission of bits across a link – modulation, error coding, interleaving and so on; they vary in performance but are not the key to the nature of WLANs.

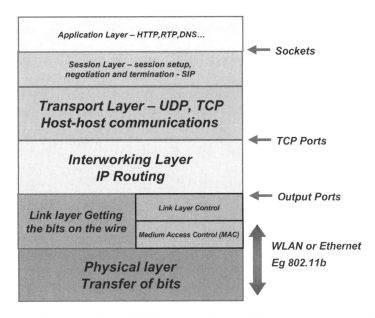

Figure 3-7.   IP stack – WLANs (IEEE) only specify the MAC and Physical layers. (*Source:* Author. Reproduced by permission of © British Telecommunications plc.)

3.4.1 MAC Layer

The MAC layer is the most important aspect of WLAN behaviour – controlling how:

- Different terminals gain access to the radio to transmit;

- Lost packets are discovered and dealt with;

- Terminals discover access points and register/authenticate with them;

- IP packets are fragmented;

- Terminals are addressed and identified;

- Data is encrypted and QoS is (or not) implemented.

In WLANs the link layer control (LLC) has minimal functionality and the MAC more or less interfaces to the IP layer directly. As we will see there can be multiple WLAN hops and access points can be connected by Ethernet (or WLAN) and offer layer 2 handover. However the basic model is that the IP packets are accepted by the MAC layer for transmission and then appear at the destination terminal/access point.

The most important function of the MAC is the sharing of the radio channel between competing stations (i.e. terminals and access points). The simplest sharing scheme (for radio spectrum or a common wire, say) was invented in

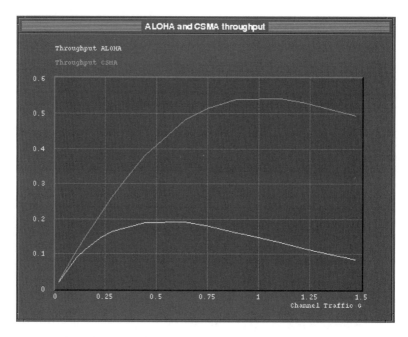

Figure 3-8.   Throughput as a fraction of the maximum network capacity versus traffic load (fraction of maximum capacity) for ALOHA (lower trace) and CSMA (upper trace). (Source: Author. Reproduced by permission of © Dave Wisely, British Telecommunications plc.)

Hawaii in 1971 when scientists (mad or not) were offered a small amount of spectrum and wanted to use it to pass packet messages between a number of islands. They discovered that the simplest way to share the spectrum was to transmit immediately when you had a packet ready to go. If two stations transmitted at the same time then a packet collision occurred and the packet was lost but this was catered for with a positive acknowledgement (ACK) sent back from the receiver to the transmitter for every correctly received packet – if the ACK was missing then the transmitting station simply sent the packet again. Figure 3.8 shows the throughput of this ALOHA protocol as a function of the network load. As would be guessed it works very efficiently at low traffic loads – when collisions are rare, but falls off as the load increases.

Far better at high loads is a scheme called Carrier Sense Multiple Access (CSMA) – which essentially says that a station with a packet to transmit must first test to see if the medium (radio spectrum or wire, say) is being used: if not then it can transmit immediately and if it is busy then the station must wait until it is free. Figure 3.8 shows the great improvement that this brings at higher loadings. Collisions still occur – when a station starts to transmit it takes a finite time for this transmission to propagate to other stations. If one of these stations tests the medium during this time it will find it free and start to transmit – and a collision will occur. The "cost" – in terms of lost capacity – of a collision depends on the time wasted transmitting during the collision. As an example the maximum size

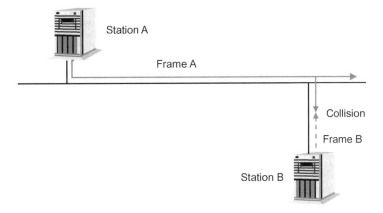

Figure 3-9. Ethernet collision detection. (*Source:* Author. Reproduced by permission of © Dave Wisely, British Telecommunications plc.)

of WLAN packet is about 12 000 bits – at 10 Mbit/s that is 1.2 ms lost – so 50 collisions a second cost 60 ms or 6% of the maximum capacity. Figure 3.8 shows that the maximum throughput – with a number of stations and random traffic – is about 60% of the capacity achievable between two isolated stations transmitting one to one.

In wired Ethernet there are two further refinements. Firstly, transmitting stations monitor the wire to determine if a collision is taking place by testing if the signal on the wire corresponds to what they are transmitting. If a collision is detected then the transmission of the packet is stopped immediately and a "jabber" signal sent to indicate a collision (see Figure 3.9). This procedure can save a lot of bandwidth – collisions can be detected and signalled in only a fraction of the transmission time of maximum length packets. The second refinement in Ethernet is that when a collision occurs then the stations colliding execute a back-off procedure – essentially rolling dice to determine the length of time they have to wait to re-transmit. In Ethernet this is called binary-exponential back-off: After a collision stations randomly back-off from 1 to $(2^n - 1)$ slot times (512 bits) with $n = 1$ initially and increased to $n = 9$ if further collisions occur up to a maximum wait of $2^{10} - 1$ (53 ms). Six more collisions are allowed (16 total) and then the frame is dropped. This scheme is called CSMA/CD (Carrier Sense Multiple Access with Collision Detection). Figure 3.10 shows that the through-put of CSMA/CD can reach 90% under heavy load and is a great improvement on CSMA because, although collisions still occur, they are less costly.

WLANs – and other radio systems – however, are not able to use CSMA/CD. On a wire the signal from a remote station is pretty much the same strength as it was when it was transmitted and collisions can easily be detected. In a radio system the received signal from a remote station is typically 60 to 80 dB down (i.e. 6 to 8 orders of magnitude smaller) than your own transmitting power. If you are listening to a channel as you transmit you will never detect such a faint signal against the back drop of your own

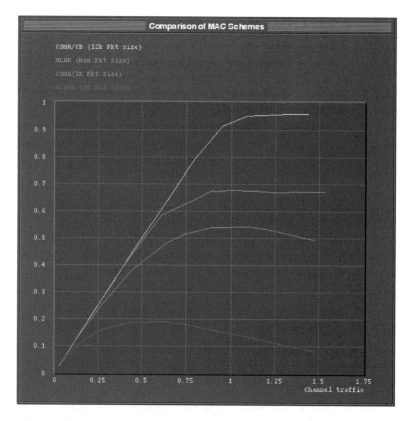

Figure 3-10. Throughput of: CSMA/CD (Ethernet) (Top trace); Hiperlan (a WLAN-second trace); CSMA (Third trace) and Aloha (bottom trace). (Source: Author. Reproduced by permission of © Dave Wisely, British Telecommunications plc.)

transmission. However, a station between you and the remote transmitter will hear both signals more or less equally and a collision will occur. The solution for WLANs is a scheme called CSMA/CA – Carrier Sense Multiple Access with Collision Avoidance.

In CSMA/CA (Figure 3.11) a station is able to transmit if the medium is free for more than a specified time – otherwise it must wait for the medium to become free (the CSMA bit). After this it must wait a random amount of time (the back-off) period – which will be (more than likely) different for different stations waiting to transmit (otherwise they all pile in when the medium is free and cause a multiway collision) – this is the CA part. If two stations – by random chance – chose the same back-off then there is a collision. This is detected by the lack of a positive ACK from the receiving station – then a larger back-off maximum is chosen.

For 802.11 WLANs the actual MAC used is called the Distributed Control Function (there is actually another one – the Point Control Function- which is specified but has not been implemented). Figure 3.12 shows DCF in more

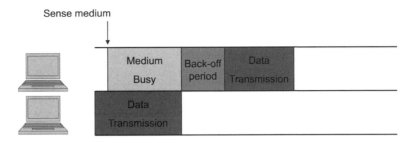

Figure 3-11. Carrier sense multiple access with collision avoidance (CSMA/CA) – note the ACK is not shown (see Figure 3.12 for the detailed view!). (Source: Author. Reproduced by permission of © Dave Wisely, British Telecommunications plc.)

detail. The wait period before the medium is deemed free is called the DIFS (DCF Inter Frame Space) and the back-off time is given by:

- Back-off time = [0, CW] * SLOT;
- Collision Window – CW – varies from CW_{min} to CW_{max} (Integer values);
- After a successful transmission CW is reset to CW_{min};
- After a collision $CW_{new} = 2*(CW_{old} + 1) - 1$.

If, during a back-off, the medium becomes busy then the station suspends the back-off and waits for the medium to become free. It then continues the back-off from the value it had reached when it was suspended. The ACK is sent after a time SIFS – the Short Interframe Space. This ensures that the ACK is transmitted before any station waiting with data. The different parameters of the MAC layer are different for different PHY layers – as an example these are the settings for 802.11b:

- DIFS = 50 μs;
- SIFS = 10 μs;
- $CW_{min} = 31$, $CW_{max} = 1023$;
- Min back-off = 670 μs, max back-off 20 ms;
- SLOT = 20 μs.

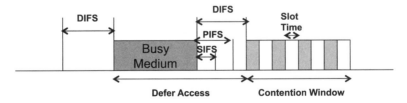

Figure 3-12. Detailed view of CSMA/CA as implemented in the DCF (Distributed Control Function) in 802.11 WLANs. (Source: Author. Reproduced by permission of © Dave Wisely, British Telecommunications plc.)

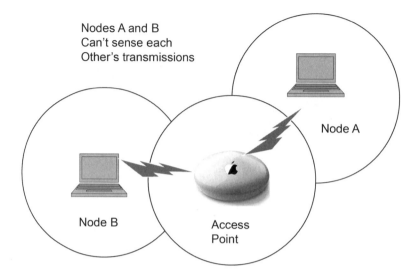

Figure 3-13. *Hidden node problem in WLANs.* (*Source:* Author. Reproduced by permission of © Dave Wisely, British Telecommunications plc.)

There is one further refinement that is needed in WLANs (and more generally in radio systems) to tackle the (dreaded) "hidden node" problem. Figure 3.13 shows how this comes about – Node A is trying to transmit to the access point and listens carefully to see if the medium is free (CSMA) it is too far from B to detect B's transmission and so – even if B is merrily transmitting (to the access point or anyone else for that matter) – Node A senses the radio channel is free and transmits. At the access point, however, transmissions from both A and B interfere and Node A never receives an ACK for the packet sent. Node A then assumes a collision and listens again to the medium – which it again senses as free and transmits again ... as you can imagine, with many nodes all with varying degrees of radio connectivity, the result is chaotic, packets are lost and the CSMA/CA performs badly.

The solution adopted by 802.11 MACs is shown in Figure 3.14. Instead of just sending the data packet, Node A sends a special management packet – Request To Send (RTS). This is specifically addressed to the access point and gives the size of the data packet that A intends to send If the access point senses the medium is free (i.e. B is not transmitting) it then returns a Clear to Send. Cunningly this is sent after the short interval SIFS – to pre-empt any stations hearing the RTS and then starting to transmit because they think the medium is clear – remember they must wait a minimum of DIFS (> SIFS). The CTS contains information about the packet length A is about to send and is heard by B (since it is sent by the AP) and causes B to update it Network Allocation Vector (NAV – a sort of map of when the radio channel is being used) which causes it to wait as if the medium was busy.

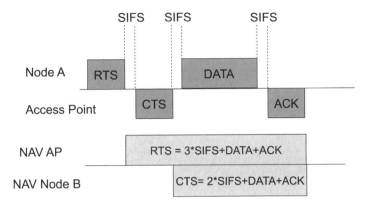

Figure 3-14. RTS/CTS used to avoid the hidden node problem. (Source: Author. Reproduced by permission of © Dave Wisely, British Telecommunications plc.)

The good news is that all 802.11 systems use exactly the same MAC[1] – the bad news is that the MAC is so entrenched, so cheaply optimised in silicon, that it has proved very resistant to change and evolution. As we shall see below the MAC is simple and robust but does not offer security, QoS or support power saving. It also gives very poor performance with VoIP due to packet overheads and can perform badly in dense radio environments where multiple systems are in operation. However, when you are struggling to book those tickets to Alaska at 10 kbit/s at the bottom of your garden and thinking about suing PC World for selling you a system with 54 Mbit/s on the label spare a thought for the original designers of the 802.11 system in the 1980s. All they wanted to do was create a wireless version of Ethernet. There was no need for QoS, security, handover or power saving as the "target scenario" was a small office with fixed PCs – hardly anyone had laptops in those days and VoIP had not been thought of. For its original intention the 802.11 MAC is an excellent solution. For a wireless city trying to emulate a cellular mobile system the MAC lacks functionality. But before we can delve deeper into these shortcomings and their possible solution there is a bit more work to do on MAC and then a consideration of the PHY layer.

A typical application of WLANs is where an access point allows a number of terminals to access a fixed network infrastructure – such as in the home for Internet access or office for Intranet access. In this case (called Infrastructure mode) the AP has a special status and it, along with all the terminals communicating with it are called a Basic Service Set (BSS). In a BSS the access point broadcasts the name of the WLAN – the Service Set Identifier

---

[1] Not strictly true – 802.11a does not have backwards compatibility.

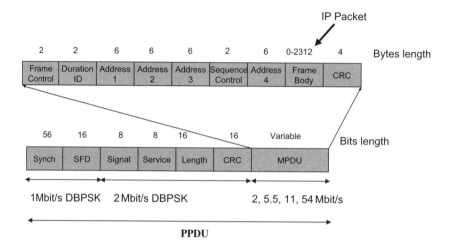

Figure 3-15. 802.11 MAC data frame. (Source: Author. Reproduced by permission of
© Dave Wisely, British Telecommunications plc.)

(SSID). When a terminal ("terminals" covers a multitude of devices such as
Laptops, PDAs and WiFi phones but they are referred to as "stations" in the
standards – this strikes me as a bit archaic) is first switched on or comes in range
of an AP then its WLAN client goes through an authentication procedure:

- WLAN client listens for access points within range.
- Client searches for APs with SSIDs matching those set in driver software.
- Client sends an authentication request to the AP.
- AP authenticates the terminal.
- Client sends association request to AP.
- AP associates with the station.
- Terminal can now access the Ethernet that the AP is attached to.

There is more about authentication and security later in the chapter but it is
important to look at MAC functions over and above the transmission of packets.

Firstly, it is responsible for addressing – every WLAN (and indeed fixed
Network Interface Card) has a globally unique 48 bit address assigned (burnt
in the silicon) when it is manufactured. In 802.11 there is provision for
stations to relay packets at layer 2 – i.e. the next layer 2 hop might not be the
final layer 2 destination. This leads to the presence of 4 48 bit addresses in
every data frame[2] (Figure 3.15). The other important point to take from

---

[2] The need for four addresses arises if you have a layer 2 network (e.g. a mesh). There are the original
and final destinations but in addition you need the addresses of the next hop and of the last hop. In
practice WLAN meshes are constructed, addressed and routed at the IP layer but the capability of
doing this at layer 2 is retained in the MAC.

Figure 3.15 – although we are jumping the gun a bit – is that the header of data frames is sent at a lower rate (1 or 2 Mbit/s) than the MPDU (MAC Protocol Data Unit) (up to 54Mbit/s). As we shall see later this leads to a situation where these packets cause the medium to be sensed busy much further than data can be transmitted.

There are many other types of frame – for control there are RTS/CTS and ACK that were introduced above and also for management such as beacon, authentication and association frames.

3.4.2 WLANs – History and the Physical (PHY) Layer and IEEE 802.11 Groups

WLANS are TDD (time division duplex) systems – using a single chunk of spectrum for uplink and downlink (Access Point to terminal) transmissions. In fact – there is no real concept of up and downlink – as far as MAC transmission is concerned – the access point has no special or privileged status to other users on the same channel. Compare this to base-stations in cellular systems that are in complete control of radio resources performing admission control and only allowing transmission opportunities (e.g. time slots) to terminals when the required QoS levels can be met. As the description of the MAC above shows WLANs use Time Division Multiple Access (TDMA) to share access to the radio spectrum between different terminals.

Worldwide 802.11 WLANs are specified to operate in the 2.4 GHz ISM Band (Industrial Scientific and Medical) – this is 75 MHz of spectrum that can be used for unlicensed radio systems that conform to a number of rules, the most important of which is that maximum transmit power is typically100 mW (20 dBm) in most countries. This is the reason that WLANs have such a feeble range when compared with cellular systems – 3G base-stations, for example, put out 60 W or so and 3G phones up to 1 W – and 3G receivers are up to five times more sensitive than their WLAN equivalents. As far as 802.11 WLANs are concerned each channel occupies 20 MHz so there is room for only three independent channels. However, there are actually 13 (11 in North America) channels specified but these overlap considerably – the section below looks at the level of interference for systems on adjacent channels – this is shown in Figure 3.16.

In the original WLAN specification – known as 802.11 (with no letter following – as explained below) used frequency hopping to spread a 1MB/s data rate packet within the 20MHz channel or a Direct Sequence Spread Spectrum (DSSS) for 2 Mbit/s. The IEEE released the 802.11 standard in 1997 but most equipment from different manufacturers was incompatible until the formation of WECA in 1999 and the start of WiFi compatibility testing.

802.11 was superseded in 1999 by 802.11b – higher rate WLAN – which offered rates of 1, 2, 5.5 and 11 Mbits rates in data packets (but not in the headers which were fixed at 1 or 2 Mbits/s as were the control packets – as shown in Figure 3.15) using Direct Sequence Spread Spectrum techniques. Note this is just a way of spreading the signal over a full 20MHz – at 1 Mbit/s it would

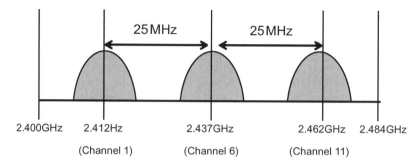

Figure 3-16. *The ISM band allows only three independent 802.11 WLAN channels.* (*Source:* Author. Reproduced by permission of © Dave Wisely, British Telecommunications plc.)

occupy 1 MHz or so without spreading. The point of using the full channel is to give much better resistance to external interference. In the case of the ISM band interference might be from other, non WLAN systems, remote WLANs on adjacent channels or microwave ovens – which operate at 2.4 GHz. Without frequency hopping or DSSS real world system performance would be unpredictable.

More recently – and certainly if you have bought a WLAN laptop or access point in the last three years or so – you will have an 802.11g system (in North America a very similar system is called 802.11a – this is described below but offers the same data rate).

802.11g operates in the ISM band on the same channels as 802.11b and offers a much higher maximum data rate – 54 Mbit/s headline rate – using OFDM (Orthogonal Frequency Division Multiplexing – this will be described in Chapter 5). OFDM is well suited to WLANs – as we noted above a WLAN channel is 20 MHz wide but unpredictable spectral components of this can have very heavy interference or be in deep fade. OFDM has the merit of dividing up the channel into hundreds of sub-carriers and is able (through coding or link adaptation) to effectively only use those that have good propagation characteristics and ignore those that are in fade or are suffering interference.

WLANs are also specified to operate in unlicensed frequencies between 5 and 6 GHz around the world. In North America the 802.11a system uses OFDM in 20 MHz channels to offer rates (depending on signal and interference) of: 54, 48, 36, 24, 18, 12, 9 and 6 Mbit/s. 802.11a has 12/13 nonoverlapping channels but this is being extended by many countries into the B band (see below) of 5.47 to 5.725 GHz – this will add another 12 nonoverlapping channels – i.e. 8 times the capacity available in the crowded ISM band.

In Europe the 5GHZ is complicated – essentially there are three different bands each with different power limits and regulations.

- 5150–5350 MHz Band A
  −60 mW indoor use (200 mW with DFS and TPC)
- 5470–5725 MHz Band B
  −200 Mw for indoor use – 802.11a permitted
- 5625–5725 MHz
  −1 W indoor and outdoor use (with DFS and TPC)
- (5725-5875MHz) Band C
  − Fixed Wireless Access
  − Light regulation

In two of the bands Dynamic Frequency Selection (DFS) and Transmitter Power Control (TPC) are required. DFS means that the terminal or access point has to scan for other users (typically military radars or other non WLANs) and then choose an unused channel. TPC requires WLANs to transmit with reduced power when they are close together or radio conditions are good – to reduce the overall level of interference to other (non WLAN) users of the bands. ISM band and 802.11a systems have no power control – they continually operate at full power – something that contributes to the poor battery life of WLANs phones and laptops. In Europe and Japan 5 GHz systems with DFS and TPC are specified as 802.11h. Unfortunately, these systems have been under discussion for five years and – although implemented in some chip sets – are basically unused today. Some of the reason is the extra cost involved but mainly it has been the lack of perceived need for anything other than ISM systems.

Table 3.2 provides a comparison of WLAN types to round off this section. Also included are Bluetooth (which is not a WLAN) and Hiperlan/2 (Ref. 9) which was an advanced WLAN standardised by ETSI in the late 1990s – it was very advanced with QoS and other features but was a commercial failure due to the rise of 802.11 WiFi products.

WLAN	802.11a	802.11b	802.11g	Hiperlan/2	Bluetooth
Carrier	5 GHz	2.4 GHz	2.4 GHz	5 GHz	2.4 GHz
Channel Bandwidth	20 MHz	22 MHz	22 MHz	20 MHz	1 MHz per Channel
Max Physical rate	Up to 54 Mbit/s	Up to 11 Mbit/s	Up to 54 Mbit/s	72 Mbit/s	1 Mbit/s
MAC	CSMA/CA RTS/CTS	CSMA/CA RTS/CTS	CSMA/CA RTS/CTS	TDMA/TDD	TDMA/TDD
Frequency Selection	OFDM	DSSS	OFDM	OFDM	FHSS
Modulation	BPSK QPSK QAM	QPSK	BPSK QPSK QAM	BPSK/QPSK 16 QAM	GFSK
Cell Radius (Typical)	50 m	100 m	40 m	30 m	10 m

Table 3-2. Comparison of WLANs.

3.5 Performance of WLANs

As an example of how careful you need to be in looking at headline rates for WLAN throughput it is, perhaps, useful to go through an example of real-world performance. Suppose you were an IT manager for a bank and the finance director said he had read an article about voice over the Internet and almost fell off his chair when he compared the cost of mobile phone usage on the dealing floor with Internet prices. He has decided the trading floor will use Voice over IP and has already purchased 100 WiFi only phones and a low-cost account at Skype – he wants you to install a WLAN to support this. So you go to the local branch of PC World and have a look at what is on offer – the bank has a high speed wired Ethernet and so all you need are some access points. A typical 802.11 b access point costs £30 and claims to offer 11 Mbit/s and a range of 30 m. Is 11 Mbit/s enough? Well suppose half the dealers are using the phone at one time – and even fixed phones only use 20 Kbit/s – that means the maximum throughput you'll need is 1 Mbit/s (50 phone calls at 20 Kbit/s) – so even with 1 Mbit/s each way you'll have plenty of bandwidth to spare. What do you think – will it work?

Figure 3.17 might be a useful guide at this point – it shows how the throughput of 802.11b varies with the size of the packet being sent for two isolated stations

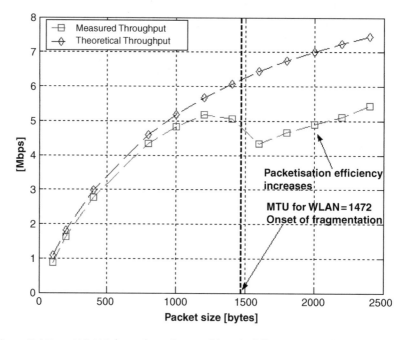

Figure 3-17.   – WLAN throughput (802.11b) with different sizes of IP packet load – measured (lower trace) and simple theory (upper trace). (Source: BT. Reproduced by permission of © British Telecommunications plc.)

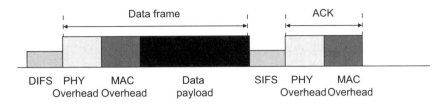

Figure 3-18. 802.11 Frame structure (with RTS/CTS turned off). (Source: Author. Reproduced by permission of © Dave Wisely, British Telecommunications plc.)

with no interference and traffic in one direction only (i.e. ideal conditions). The first thing to notice is that the maximum throughput – as measured by a real application (i.e. not including TCP, IP and MAC layers headers or PHY layer error coding) is only 6 Mbit/s. The next most striking point is the very alarming fall-off of throughput as the packet size is decreased – falling to under 1 Mbit/s for packets of 100bytes (800 bits). Finally notice that the measured throughput suffers a dip at 1472 bytes (11,776 Bits) – the maximum size of IP packet that can be carried in MAC frame. Bigger IP packets are broken up in to a maximum size packet and a small "off cut" that is sent in a small frame by itself – this explains the measured throughput efficiency falling in Figure 3.17 for packets over 1500 bytes.

So what causes this precipitous loss of efficiency for small packets? In a word – overheads! When an IP packet is transmitted it must be:

- Fragmented if it is over 1472 bytes long;

- Formed into a MAC data frame (see Figure 3.15) – which contains 366 bits of other information that is not part of the IP packets (address, versions type, synch signal etc.);

- Formed into a PHY layer frame – with more overheads including error coding bits.

When the PHY frame is ready to go the station must wait the minimum time to see if the medium is free (DIFS)[3] – it then transmits the frame and waits for the receiver to send back the ACK (which will only happen after the shorter inter frame gap SIFS). Figure 3.18 shows the whole sequence and how the actual time spent sending data can be quite small. Coming back to the example of VoIP the key network parameter is the total delay – the time from me saying "hello" to the complaints department at Network Rail hearing it.

---

[3] From what was said earlier you might think the transmitting WLAN could just send the packet if the channel was deemed free – this is true but after a successful transmission the terminal must then wait DIFS plus a random back-off before using the channel again – this is called the "post-back-off" and ensures there is at least one back-off gap between data frames – it also allows other terminals a fair chance to access the channel.

The longer the delay the lower the quality of the voice (think of the old satellite links for phone calls where you almost had to say "over" to indicate that you had finished talking!) Low levels of packet loss, especially with advanced voice coding, are much less important. Variations in delay, called jitter, can be taken up with buffers – but again these add to the total delay. Typically a maximum delay budget for voice is 100–150 ms – before the quality is perceived to fall off alarmingly (by 400 ms you need to say "over"). Many things contribute to delays but one contribution is from the packetisation delay. This is the time it takes to collect the data into a packet – the voice bits arrive in a constant stream but need to go out in packets and so the first bit must wait for this packetisation delay until a packet's worth of bits have been produced. There is a trade-off with packet size – the larger the packet the more efficiently it will be sent over the WLAN, but the longer the packetisation delay will be. Allowing 20 ms at each end for packetisation is generous and, for a speech codec with a rate of 20 kbit/s (say), a 20 ms packetisation time yields packets 400 bits long. At this size the throughput will be much less than 1 Mbit/s for 802.11 b systems and if you installed one access point in the bank you may well be looking for another job. In fact most real-life systems manage three calls on 802.11 b (such as BT Fusion – which adds security overheads as well) or eight to 10 calls on 802.11 g. The next section looks at a number of potential solutions for improving voice efficiency over WLANs but it is fair to say that today there is no standardised solution that can be bought off the shelf and that interoperates between manufacturers.

3.6 WLAN Network Coverage

The next area that needs consideration is coverage – how far can you expect it to reach? Figure 3.19 shows some simple tests made in a typical office environment.

Figure 3.19 represents the absolute maximum throughput and was made with two isolated stations transmitting in an interference free area. This and other measurements (Ref. 10) show that a number of generalities on WLAN coverage can be concluded:

Throughput is less than half the headline rate for all 802.11 systems – even in ideal conditions:

- 802.11a and 802.11g coverage falls off much more rapidly with distance than 802.11b.

- Solid walls tend to cause a significant (10 dB or so) loss.

- 5 GHz systems have the shortest range and fastest fall-off.

- There are "dead spots" in many hot spots – caused by local shadowing (e.g. a metal filing cabinet).

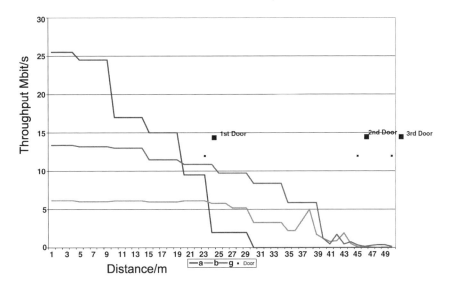

Figure 3-19. Throughput vs distance (m) in a typical office environment. 802.11a (top trace), 802.11g (middle trace) and 802.11b (lower trace). (Source: BT. Reproduced by permission of © British Telecommunications plc.)

In general WLANs can reach about 30 m indoors but manage much better outdoors with 100 m or more possible in ideal conditions. However, the range of a WLAN is comparable to a 3G Femtocell (see Chapter 4) – the latest in a line of ever smaller base-stations from macro to micro to pico (hotel) to nano (small office) and finally femtocells that are designed to cover a house. Femtocells emit about 20 mW, compared to 50–100 mW of WLANs for a similar range – reflecting the more sensitive receivers of cellular phones. As was noted earlier, cellular base-stations put out 60 W or so and that it is the main reason why cellular systems have a significantly greater coverage.

WLAN can also be used for longer range, point to point, links. These can cover over 1 km or so by adding directional antennae to the transmitter and receiver. However, it is very easy to exceed the EIRP – Equivalent Isotropic Radiated Power – limit for WLAN – essentially if you transmit 10 mW in 1 srad[4] then the EIRP is about 120 mW (i.e. 10 mW *4π) and over the limit of 100 mW.

Interestingly there was a study in the UK (Ref. 11) about increasing the power of WLANs. Powers of up to 10 W were considered – using both modelling and trials (with suitable licences from the Radio Communications Agency!). The results for an urban area with a mix of public and private WLANs in homes and offices showed that in areas where WLANs were sparse there was an advantage to increase power to 600 mW and that resulted in an overall increase in

---

[4] Steradian – measure of solid angle. Think of a football – divide the area into about 12 equal areas – maybe you have one of those balls with the black and white hexagons on? – each 1/12 of the area subtends a solid angle of 1 srad at the centre of the football.

coverage and throughput. However, when there was a high density of WLANs – 10 to 25% of houses – then increasing the power only increased the interference and reduced throughput and coverage. The problem was the lack of independent channels (3) and that WLANs were deployed without any frequency (channel) planning. This topic is picked up again in the section on interference below.

3.7 Improving WLANs

In the IEEE there are a great many working groups looking at WLAN – this is the largest of the IEEE standards area and has more participants than any other technology. Many groups have completed producing a standard (such as 802.11a, 802.11b and 802.11 g) – but many are still active and these are listed in Table 3.3. In the next section a subset of the whole "alphabet soup" of 802.11 will be discussed. Some of the completed standards – such as 802.11 h the Japanese/European operation at 5 GHz – have been completed but may never become productised. Others, such as 802.11e on WLAN QoS have been (partially) endorsed by the WiFi Alliance and are beginning to make their way into chip sets. Still others – such as 802.11n high rate WLAN using multiple antennae – are not yet standardised but have given rise to so called "pre-standards" products – these offer performance advantages but are not interoperable. Here the discussion will concentrate on groups looking at enhancing WLANs for: Security, QoS, higher speed, interference mitigation and adding cellular-like capabilities (paging, idle mode and handover).

- 802.11k – Radio Resource Measurements;
- 802.11m-802.11 Accumulated Maintenance Changes;
- 802.11n – High throughput wireless systems beyond current 802.11b/g/a systems;
- 802.11p – Wireless Access for Vehicular Environments;
- 802.11r – Fast Roaming;
- 802.11s – Mesh Radio;
- 802.11t – Wireless Performance Prediction;
- 802.11u – Wireless Inter-working with External Networks (e.g. 3GPP/Wi-MAX/Fixed etc.);
- 802.11v – Wireless Network Management;
- 802.11w – Protected Management Frames;
- 802.11y – Contention Based Protocols.

Table 3-3. Current 802.11 Task Groups (TGs).

3.8 Security

There are two main types of WLAN access points on offer – firstly, all-in-one units that act as DSL modems and IP routers and offer services such as DHCP as well as WLAN connectivity. Secondly, are more limited access points – with no modem and which can be plugged into existing Ethernets. Straight out of the box most WLAN access points can be plugged into an Ethernet or DSL and they offer absolutely no security whatsoever.[5] What does "no security" mean in the context of WLANs? Well firstly it means that anyone with a WLAN-enabled laptop can connect to the access point – they are effectively connected to your network in exactly the same way as if they had connected with an Ethernet cable. Given that a WLAN access point in a typical house can be accessed from neighbouring houses and that specialised equipment (enhanced receivers and transmitters) can be used to access enterprise WLANs from beyond security fences that can be a serious threat. Secondly, it means that even if hackers don't want to connect to your network they can read all the traffic on the WLAN. That might include password to websites, the contents of emails and so on – basically any traffic that is not encrypted by some higher layer – like secure sockets in web browsers.

A "friend" of mine used to connect to his neighbour's WLANs. He could see three WLANs from his lounge, two of which were completely unprotected. On one network he was able to access the Internet, print documents and look in shared folders on a computer connected to the home network – you have been warned! This is an offence and somebody in the UK was recently prosecuted for this.[6]

At BT Adastral Park – a site of about a square mile with many buildings and 4000 employees – when WLANs first appeared people were plugging them into the Ethernet backbone and creating a serious security issue. BT security soon moved in and were then seen in the following months with monitoring equipment trying to detect unauthorised WLANs onsite and walking along the perimeter of the campus. The problem was eventually solved by introducing an official WLAN (with strong security) and draconian penalties for plugging in anything else.

There are a number of ways to protect WLANs which offer increasing levels of security. The first possibility is to hide the SSID – the WLAN identifier (BT Openzone, Dave's House, Linksys (default) etc.) – that is broadcast by the access point in beacon frames. Terminals can only authenticate with an AP if they know the SSID but this is basically useless as it can be found using suitable tools. One of the best tools for detecting WLANs is Netstumbler (Ref. 12) – which is capable of providing a great deal of information on WLANs in the vicinity.

---

[5] This is becoming less common in 2008. For example most home hubs (WLAN router-modems) supplied by ISPs (e.g. BT) all have security as standard.

[6] Two people were arrested and cautioned for this in the UK in 2007 – see http://news.bbc.co.uk/1/hi/england/hereford/worcs/6565079.stm.

The next level of security – in terms of complexity and usefulness is MAC address filtering. You will recall that every WLAN chipset has a fixed, globally unique, 48 bit MAC address. Most access points offer the ability to filter a set list – either a white list (allowed) or a black list (banned). This can be useful to exclude casual hackers – I have used this where a single access point had to serve a variety of terminals (MAC, Windows XP; Old Windows 2K laptop with PCMCIA card; Wii; WLAN camera) – I simply could not get higher security (like WEP – see below) to work with all these various devices. However, I wouldn't recommend it if you need serious security as MAC address spoofing is relatively simple. There are many utilities that can be used to set the MAC address sent out by various WLAN chip sets to any desired value. All a hacker would need to do is to passively identify the MAC address of a station on the white list and use that to gain access to the WLAN. Also there is no encryption of the traffic and no mechanism to prevent malicious hackers from injecting traffic. This is far from an academic point – imagine you are in an auction (on line or on the phone) and at the moment of denouement – as the hammer comes down – a hacker send a de-register message to the access point – that is that – your connection terminated for at least two minutes!

The first "serious" WLAN security technology is WEP – Wired Equivalent Privacy (Figure 3.20). Introduced in 1999 WEP uses secret keys – manually entered in all terminals and access points – of length 40 or 104 bits (known as WEP2). WEP works by combining the key with an Initialisation Vector (IV) (24 bits) to generate a bit sequence as long as the data to be protected and XORing the two to produce an encoded payload (this is called a stream cipher and uses an algorithm called RC4 – Ref. 13). In theory WEP allows only terminals that have the key to authenticate with access points, encrypts data traffic and prevents injection of packets.

However, WEP is seriously flawed. WEP can be passively cracked in a few hours by anyone able to capture all the traffic on the network – even a modest laptop and suitable software "available on the Internet for free" will do. My "friend" referred to above was able to crack the WEP key from a well-known ISP's standard issue WLAN router after three days continuous recording of his neighbour's WLAN – the amount of time varies with

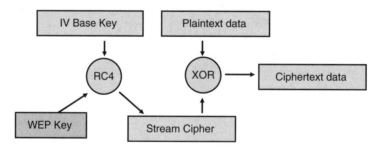

Figure 3-20. WEP key encryption process. (*Source:* Author. Reproduced by permission of © Dave Wisely, British Telecommunications plc.)

network use and algorithms are known (Ref. 14) that give a 50% probability of 104 bit key recovery in a minute on an 802.11g network running at full speed!

So what is wrong with WEP? Firstly, the big issue is the use of the IV – this is sent in the clear (i.e. unencrypted) with the encrypted data. In addition there are a number of packets in which much of the content is well known and fixed (such as ARP – address Resolution Protocol – used to link MAC and IP addresses together). So the hacker has: part of the key, the unencrypted data and the encrypted data. This (slowly) allows the discovery of the hidden key. Things are speeded up greatly when the IV is reused – it is supposed to change for every frame but in practice always starts from 0 at a reset and it is also quite short and "rolls-over" very quickly on busy networks. The more IVs collected – the quicker the "crack" (some IVs are also "weak" in that they are known to lead to the key very quickly).

So WEP will always be cracked by a determined hacker – whatever key you chose – and 104 bits is not anything like 2.5 times better than 40 bits! The problem is exacerbated by the fact the WEP key is the master key and serious code-breakers know that you never use your master key directly – instead you use it to derive temporary keys and change them frequently. This is part of the solution defined by the 802.11i working group (now complete). I would not recommend WEP for enterprise use – it might be OK to protect your home network but monthly key changes and regular monitoring of traffic and connected users might be advisable. If in doubt – upgrade to the next level of security.

The 802.11i standard was complete in June 2004. Previously the WiFi Alliance had introduced a sub-set of the 802.11i standard – and this was endorsed and marketed as WPA (WiFi Protected Access). This fixed some of the issues with WEP – in particular the introduction of temporary keys but has now been superseded by WPA2 (also known as RSN – Robust Security Network) which is a full implementation of the 802.11i standard. WPA2 works in a fundamentally different way to WEP and can offer very strong network protection and encryption – as strong as cellular mobile, for example.

One of the major innovations in 802.11i is the use of three-way authentication. A protocol called 802.1x (another IEEE group) allows minimal access from the terminal to the access point to allow the terminal to authenticate itself to a remote server (Figure 3.21). A protocol – called Extensible Authentication Protocol (EAP) – is then used to transport the actual authentication protocol (of which there is a very wide choice based on: name/password; certificate; SIM card and fixed keys). Once the terminal has authenticated itself to the server (which has a pre-existing security association with the access point – i.e. they "trust" each other) then the server validates the terminal to the access point and access is allowed. The AP and server might be part of an Enterprise network, for example, or the server might belong to an ISP and the AP a network provider. This means that access points don't have to make security decisions – all the authentication and security

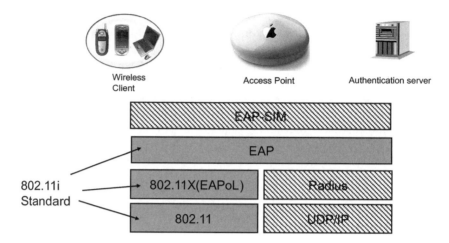

*Figure 3-21.   802.11i components in the authentication of WLAN terminals. (*Source:* Author. Reproduced by permission of © Dave Wisely, British Telecommunications plc.)*

policies etc. are centralised and can easily be changed. Strong security can also be provided in a way that each AP could not – with a SIM, for example.

At the end of the three-way exchange the terminal, access point and authentication server have established a trust relationship between themselves – with the advantage that the terminal now knows the access point is "genuine". In addition a pairwise master key is derived by both the access point and terminal (but never transmitted over the air). This key lasts for the whole of the session the terminal is associated with the AP and so, to further minimise the chance of its detection, it is used to derive a pairwise transmit key (PTK) that lasts only a limited time.

WPA2 is superior to WEP because the keys used to encrypt each frame can be longer (the master key might be short but this is not used directly – it might also be very long as in a certificate), the keys are changed frequently and a much improved encryption algorithm (AES) is used. Altogether, WPA2 is considered as strong as that in cellular systems in principle.

There are two different versions of WPA2 – an enterprise version which operates with a remote authentication server as shown above and a consumer version that incorporates the authentication function into the access point. In the consumer version a fixed key (the pairwise master key) is entered in the access point and all the terminals. WPA2 is so secure that hackers now use a dictionary attack on this key – i.e. try thousands of common words (don't use Dave!). Most APs offer a small algorithm for creating a long random key and you should always use this. With a long random key WPA2 is considered uncrackable (today at least!).

In the enterprise there are many different EAP versions that use passwords or certificates and some that are proprietary (such as Cisco's LEAP). In general WPA2 is slowly being introduced in some enterprise systems. Here at BT

Adastral Park, for instance we have two WLANs systems. One uses WPA2 but the other uses a higher layer security mechanism based on Virtual Private Networking (VPN) – these are established end-to-end and encrypt all traffic sent over them. Similarly public WLAN hot spots don't use WPA or WEP but a higher layer (browser hi-jack) technique. This is further elaborated in the next section on roaming. Most consumer access points/routers/modems now support WPA (2) – but this is not always compatible with existing clients and drivers – updates are available but many people have not installed them.

3.9 Roaming

When you go abroad today and switch on your mobile phone at your destination the chances are it will just work. In fact what is happening is that the phone is detecting a number of networks and attempts to authenticate to them in turn until one succeeds. The actual authentication is done by the home network – the MISDN (see Chapter 2) (a number unique to the SIM) – allows the visited network to identify the home network and secure signalling is established to allow the SIM application to authenticate itself. The commercial side of roaming for GSM – how records of calls are made and inter-operator payments made – is mostly handled by the GSMA (GSM Association – a club of GSM operators worldwide).

In contrast WLAN roaming is at a much earlier stage of development. I was in the Gare de Midi in Brussels the other month – killing time until my train after an EU project audit – and wanted to check my email. Now I have business and home BT Openzone accounts and vouchers from T-Mobile and a couple of other well-known hot spot operators. When I fired up my laptop I could see six WLANs networks – none had useful names and I had no idea if any of them was a roaming partner of a network for which I had an account or a voucher. In the end I tried them in order and the third one was unsecured and free and so I used that! That story rather sums up WLAN roaming today. The other major issue is cost – with BT Openzone roaming charges varying from 5 to 20 p/min.

Whilst operators like BT have gone to great lengths to secure roaming agreements (over 20 000 compatible hot spots worldwide at the last count) the actual mechanics and user experience has some way to go to match the GSM experience. Firstly, the way most hot spots work is to "hijack" the browser. Access points allow users to associate with them and gain an IP address but a firewall prevents any network access other than for web browsing. However, when they request their home page an application layer gateway (a web server proxy) sends back the hot spot operator's splash screen and no further access is possible until a log-in has taken place (Figure 3.22). This is the first roaming issue – the name/password format of different operators varies enormously and accommodating roaming partners on a splash screen is not always possible (in Openzone there is a drop down selection). Secondly, the name does not identify the operator – in Openzone I am 12283283787 and T-mobile dave.

Figure 3-22. BT Openzone splash screen. (Source: BT. Reproduced by permission of © British Telecommunications plc.)

wisely... The next issue for WLAN roaming is how to authenticate roaming users – currently the most common way is for the home operator to email the roaming partner a list of name/password combinations that are eligible for roaming! That just about works when the only service you are selling is Internet access but for services such as voice, video and music download there has to be the concept of a user-profile with details about services the user has subscribed to. With most commentators expecting the price of basic Internet access to continue to fall (probably to nothing!) operators increasingly need "sticky" services and value-add – things that require customer profiles, presence information and so on. Finally roaming usually entails payment of some sort – tricky today when some operators charge by the Mbyte and some by the hour. Tomorrow, however, when VoIP and video streaming are the big WLAN services (say) then WLAN roaming will need some of the subtlety of GSM roaming (call barring, roaming charges, home control, short dialling codes on roaming and so forth).

Figure 3.23 shows the WLAN roaming test-bed that the author's lab is working on which incorporates a trial of WLAN roaming using EAP back to

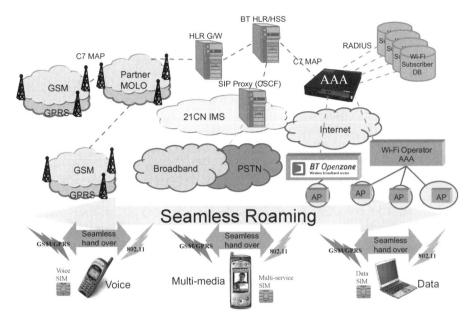

Figure 3-23.   Seamless roaming – test-bed at BT. (Source: BT. Reproduced by permission of © Dave Wisely, British Telecommunications plc.)

a AAA server here in Martlesham. The trial is part of the work of the Fixed Mobile Convergence Alliance (FMCA) and you can read more about this at http://www.thefmca.com. The user experience for WLANs in the trial is very similar to the SIM/GSM roaming experience – networks are automatically selected and authentication is carried out back to the home AAA server without user intervention. The security is strong and could be used for additional services such as e-commerce.

Looking beyond WLAN-WLAN roaming is 3G-WLAN roaming which is covered in the chapter on convergence (Chapter 6).

3.10 Quality of Service (QoS)

When talking about quality on WLAN it is useful to distinguish between scenarios. If your WLAN access point keeps crashing then you could say QoS for you is poor. Similarly if you use your WLAN to access popular websites in the evening then the response might be slow – but no slower than for a fixed connection because those sites (or the links to them) are overloaded. If you have a leaky microwave oven then you might find your wireless data rate falls when your spouse heats up the milk. These are all things that can affect QoS in the wider sense but what I want to concentrate on here is QoS in a narrow sense: the access point is working, there is no interference, but there is (say) 50% more traffic than the WLAN can take – or a mixture of real-time (voice, video say) and nonreal-time (WWW, email, downloads etc.). That, incidentally, is the big issue with IP QoS in general. If the network is only loaded 50% or less then you don't really need QoS – you are effectively over-provisioning. When the network is loaded from 50% to (about) 200% then a QoS scheme is useful – it can differentiate between users and types of traffic and can give priority to real-time traffic. When the network is more than 200% loaded then it is just overloaded and no QoS scheme really helps – what you need is more bandwidth! Of course load is not static – networks tend to be loaded only for short periods of the day. (A bit like driving into Cambridge – there is a QoS scheme – it's called bus lanes and traffic light priorities (actually I have heard it called quite a few other things!!). In truth it doesn't work because outside the rush hour you can drive around the city without any real delay but at rush hour the whole place locks dead and nothing moves – the only solution being to increase capacity which is politically unacceptable!).[7] Much the same has happened to the Internet with many applications working quite happily without QoS most of the time but also periods when the network is unusable. This is not helped by the fact that capacity has been doubling every 18 months.

So, given this slightly negative view of IP QoS, what is the role of WLAN QoS? Well I think there are a number of reasons why a WLAN QoS scheme would be useful. Firstly, WLANs are not really used for real-time services (particularly voice) today and these are the services that suffer most when WLANs are overloaded (although that is changing). Secondly, in general, the wireless link is the bottleneck in a network. Fixed network capacity tends to double every 18 months (it pretty much follows Moore's law). Wireless spectrum usage efficiency tends to double about every five to seven years – so I think the wireless hop will always lag behind (remember that even 802.11g only offers 5Mbit/s total throughput with mixed traffic). It is also true that protocols such as TCP perform very badly over wireless links – lost or delayed packets can signal to TCP that congestion is taking place and cause flows to slow down. Wireless links – with unpredictable losses and delays to packets – can cause TCP to assume network congestion is taking place even when the wireless link is well under full capacity.

Most QoS scheme must involve changes to the MAC and PHY layers to provide differentiation between different types of traffic or different users.[8] If an end-to-end solution is needed then an IP layer QoS solution (such as DiffServ or IntServ – Ref. 15) would be needed as well as a standard mechanism for mapping the IP layer QoS to the MAC layer mechanisms for the wireless hop – Figure 3.24 shows WLAN QoS as a component of a complete solution – readers wanting more detail might consult Ref. 15.

---

[7] To be fair not everyone agrees with this "petrol-head" point of view!

[8] It is possible to treat each AP as an IP router and provide an IP layer solution. However, this would not be subject to IEEE standards (they only cover the MAC and PHY layers) and would rely on all terminals to be upgraded for WLAN QoS.

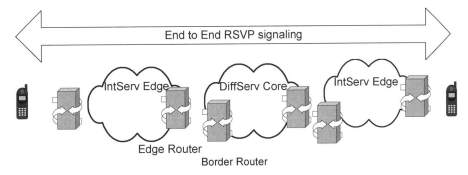

Figure 3-24. WLAN QoS solution in the context of IP QoS. (Source: BT. Reproduced by permission of © British Telecommunications plc.)

There are a number of schemes that have been developed for WLAN QoS but the most important – and the only one now likely to be used in anything other than proprietary products – is the work of the 802.11e group. This group has modified the 802.11 MAC layer to add two different sorts of QoS support – in these schemes the PHY layer is unchanged. It is obvious from the MAC description earlier that it was designed as a wireless version of Ethernet – working well when the traffic is bursty and mostly composed of "large" (i.e. max length) packets and that no applications that needed delay and jitter control were being used. If we have an access point (say) and six terminals transmitting a mixture of applications (voice, video, www, download etc.) then when the traffic gets towards 50% of the capacity of the wireless link, the delays and jitter starts to rise. This is because collisions start to happen, terminals use the back-off procedure (which is more or less random) and the sequence in which terminals get access to the medium is also pretty random – a terminal sending only voice packets might get three opportunities to transmit and then miss out on the next 20 "lotteries" for the right to transmit. In the end, of course, the opportunities to transmit are shared more or less equally but the delay and delay variation can be quite considerable in these circumstances.

In addition the throughput is not shared equally under such circumstances. Here is an example: there are five stations and an access point – four are sending maximum length packets (say 10 000 bits) and one is sending 100 bit packets (voice). If the total throughput is 5.01 Mbit/s then all stations will get 100 transmission opportunities per second but the throughput will be 10 kbit/s for the voice station and 1 Mbit/s for the others. This might not matter if 10 kbit/s is adequate for the voice traffic but if a 64 kbit/s codec was being used the quality would be very poor (a higher level application might detect this and use a lower codec – such a scheme is described in Chapter 6).

The other issue with the 802.11 MAC is that it is a FIFO (First In First Out) queue system – if a voice packet arrives at a WLAN client after three data packets then it must wait for them to be transmitted first. There is also the uplink/downlink unfairness problem – because there are lots of terminals trying to transmit up and only one base-station transmitting down. The MAC treats them all equally (they all have the same random chance of transmitting) – meaning that in a busy network with lots of uplink traffic this can get priority over downlink traffic that is coming through the access point. The solution is to make the access point special in some way – e.g. making it a master and the terminals slaves that transmit in slots allocated by the access point.

The 802.11e standard introduces a new MAC mode the Hybrid Coordination Function (HFC) – which has both contended and contention free QoS schemes. Currently only the contention-based EDCF (enhanced distributed control function) has made it in to chips sets at the present time (Q3 2007) and this part of the standard has been endorsed by the WiFi Alliance and branded Wireless Multi Media (WMM) and a number of products are available. The HCCA – HFC Controlled Channel Access – used in the contention-free periods has not yet been implemented.

EDCF introduces two new MAC ideas – packet bursting and prioritisation. When a station wins a transmission opportunity (TXOP) it wins the right to transmit for a certain time and can transmit a sequence of packets without any other station getting access to the medium for the duration of the TXOP. This is achieved by introducing a new inter-frame spacing (IFS) that is shorter than the standard frame spacing DIFS (see Figure 3.25). This is designed to allow, for example, a number of small voice packets to be sent in a single TXOP – rather than waiting for several turns.

The other aspect of EDCF is prioritisation using frame spacing and back-off window sizing. You will recall that after the medium is free the DCF MAC dictates a minimum wait of DIFS plus a random back-off between the min and max contention windows. Traffic from the higher (IP) layer arrives at the MAC layer along with a priority (0-7) – this could come from DiffServ marking on the packet, say, or from an application-layer classifier (voice = 7, video = 6 etc.). EDCF actually has four Access Categories (0 to 3 – with 3 being the highest category) to which these are mapped.

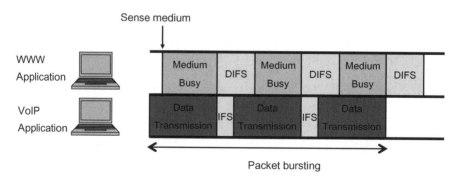

Figure 3-25. Packet bursting in EDCF of 802.11e (note the ACKs are not shown). (Source: Author. Reproduced by permission of © Dave Wisely, British Telecommunications plc.)

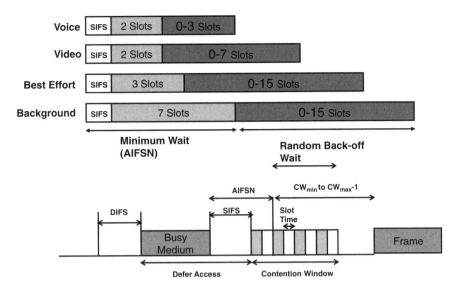

Figure 3-26.   EDCF prioritisation in 802.11e. (Source: Author. Reproduced by permission of © Dave Wisely, British Telecommunications plc.)

EDCF priorities traffic according to the AC by effectively varying DIFS and CW_{min}/CW_{max} for different traffic types (Figure 3.26). DIFS is replaced by AIFSD which is a function of AC. The actual values of AIFSD[AC] and CW_{min} [AC]/CW_{max} [AC] as well as the TXOP[AC] is decided by the AP – using a proprietary algorithm depending on the traffic level – and announced in the beacon frames. Table 3.4 shows some typical setting.

For example, voice traffic must wait SIFS plus three slots (in order that ACKs can go first) and then a random 0 to 3 slots. The only traffic that can pre-empt voice is

Priority	Access Category [AC]	Designation	AIFSD	CW_{min} (Slots)	CW_{max} (Slots)	TXOP (limit) ms
0	0	Background	SIFS + 7SLOTS	0	15	1Packet
1	0	Background	SIFS + 7SLOTS	0	15	1Packet
2	1	Best Effort	SIFS + 3SLOTS	0	15	1Packet
3	1	Best Effort	SIFS + 3SLOTS	0	15	1Packet
4	2	Video	SIFS + 2SLOTS	0	7	5
5	2	Video	SIFS + 2SLOTS	0	7	5
6	3	Voice	SIFS + 2SLOTS	0	3	3
7	3	Voice	SIFS + 2SLOTS	0	3	3
DCF and non 802.11e terminals			DIFS = SIFS + 2SLOTS	31	1023	

Table 3-4. Possible EDCF settings for different traffic classes.

video – and the odds are stacked 2 to 1 in favour of voice (0 to 3 slots compared to 0 to 7). In addition when voice traffic is transmitted it has a TXOP that allows many voice packets – further reducing jitter and delay. Normal DCF traffic – which would be used by non 802.11e equipped terminals – can be made to wait for all EDCF traffic because the CW_{min} value is typically 31 for DCF (see Table 3.4).

The AP could decide to make the DCF compete with the EDCF background traffic if by setting the AIFSD for background traffic to DIFS and CW_{min} to 31 which it is able to do – depending on traffic conditions.

Of course EDCF does not provide absolute QoS guarantees – there is still a lot of random variation in delay, jitter and throughput – although this is much reduced for real-time traffic compared to DCF. EDCF works best if there is a good mix of traffic – otherwise there is nothing to be prioritised over!

802.11e has also defined – as part of the Hybrid Coordination Function (HCF) – the HCCA (HFC Controlled Channel Access). For this part of the HCF the access point has a much greater role – more like that of a cellular base-station – at least as far as QoS in concerned. The MAC is now divided into contention and contention-free periods (where HCF operates). The contended periods are when EDCF is in operation but during the contention-free time the access point is responsible for deciding which terminals can transmit when and for how long. In this contention-free period the AP polls the terminals in turn as shown in Figure 3.27 – after a poll traffic is sent exclusively from AP to terminal and terminal to AP until the AP moves on to the next terminal. No other station (DCF or EDCF) gets a chance to transmit as the gap between the medium going

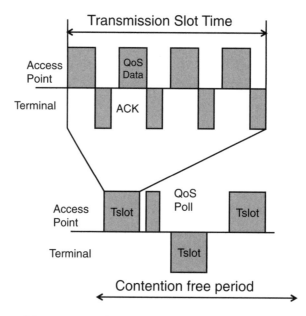

Figure 3-27.   HCCA in 802.11e. (Source: Author. Reproduced by permission of © Dave Wisely, British Telecommunications plc.)

free and poll or HCCA data frame is shorter than DIFS and AIFS (but not SIFS – to allow ACKs).

The hard part of HCCA is how the terminals describe their QoS requirements to the AP (in something called the TSPec – Ref. 16) and how the AP decides who gets what and when – the standard leaves this to the vendor to implement the algorithm for this.

At the present time (Q2 2008) no vendors have implemented the HCCA and – at least until (or if) IP QoS solutions become more widespread, there is not much prospect of it being produced. The problem – as I started to explain at the start of this section – is that WLAN QoS often works OK because there is plenty of spare bandwidth (or you can upgrade – the old over-provisioning argument) or, when QoS fails, it fails because either there is interference of some sort (which 802.11e will do nothing for) or there is so much congestion that schemes like this are overwhelmed and what is really needed is more capacity.

3.11 Handover and Paging

Handover in WLANs is not common today – in a consumer environment there is often a single access point and the same is true in most hot spots. In the wider scheme of things it is a moot point if specific handover is needed at all; if WLANs are just for Internet access then most modern operating systems (ie Windows XP/Vista) allow you to specify a number of WLAN SSIDs and enter security details. When one of these WLANs is detected the system will just connect and then if you move away from your original AP connection will be lost and the terminal will disassociate with the old AP. Eventually another AP with an SSID on your list will be found and the OS will re-connect. Applications like browsing and email will just continue – so why would you need specific handover (which might be differentiated from this sort of reconnection by being faster (50 ms as opposed to 5 sec say) and with much lower packet loss). Only for real-time applications like VoIP and video streaming would handover really be needed. If VoIP is going to be the killer application for WLANs going forward then handover will definitely be needed as Wireless Cities and extended coverage becomes common. Hand-over in WLANs is always break-before-make – meaning that the association with the old AP must be broken before a new one is established – this is to keep the radio cheap and cheerful. In cellular systems, by contrast, handover is make-before-break – with connections to both old and new base-stations simultaneously maintained.

Enterprise WLANs can engineer seamless handover (i.e. fast (10 ms) and with low packet loss) – but today this usually requires proprietary equipment from a single manufacturer. This is achieved by using a central Access Controller (AC) and what are called "thin access points". Thin access points do not have all the functionality of standard, or "fat" access points, instead functions like authentication and association is handled by the AC

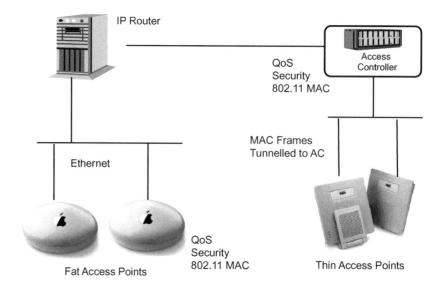

Figure 3-28.   Thin and FAT Access Point architectures. (Source: Author. Reproduced by permission of © Dave Wisely, British Telecommunications plc.)

(Figure 3.28).[9] This architecture is typically used when an enterprise needs five or more access points for coverage – the IT manager is then able to configure SSIDs, authentication, channels and all aspects of WLAN management centrally – which, if you have ever had to apply a firmware update and re-initialise 50 APs – you would realise is a great time saver. Typically these AP/AC networks offer functionality not available on single AP systems such as: multiple SSIDs – each with a different authentication mechanism (WEP, WPA, open access etc.); QoS with different traffic classes; particular terminals given priority and logging of access records. The other great advantage of AP/AC systems is that they offer fast handover – meaning authenticated terminals can be handed between APs fast enough (10 ms or so) and with low enough packet loss to allow VoIP and video streaming. To put this in perspective if you wanted to use WLAN to offer VoIP over your house and you had a largish house you might need 2 APs for coverage. If you chose these from the same manufacturers (e.g. Apple Airport is one brand we have had success with) then your VoIP application will move between access points quite happily without any perceptible glitch – you'd be hard pressed to notice the handover. Turn on security, however, and some access points (particularly between different manufacturers) will generate a 2 to 10 second gap in which no traffic is transmitted – this is fine for WWW browsing and email but usually fatal to VoIP applications. There is, you will probably not be surprised to hear, an IEEE working group – 801.11r – that is working on a fast handover solution. However, the AP/AC solution already does this – basically because terminals are authenticated only once, when they join the first AP, and then any subsequent authentication/association requests are relayed to the AC which immediately replies. There are other mechanisms built into the AC/AP architecture – the APs all monitor the strength of received signal from the terminals and report these back to the AC. Normally terminals will not disassociate from an AP until the signal to noise level is very poor indeed – even if another AP on the same channel and SSID is available. In this case the AC instructs an AP to force a terminal to disassociate – the terminal then automatically scans and finds the new AP with much better S/N ratio. In addition the AC is able re-route data for a terminal to a new AP as soon as the AP has received an initial frame from the terminal. Typically the entire AC WLAN zone is configured as a single IP subnet – so terminals keep the same IP address throughout the zone – ACs often offer DHCP functionality for this purpose – as well as NAT (Network Address Translation) and firewall functions if needed.

There are two major drawbacks with AC systems – firstly, they are relatively expensive only being suitable when five or more APs are needed and, secondly, they are proprietary and you need to buy the AC and APs from a single manufacturer.

This will change when the IETF CAPWAP protocol for access point control is completed and adopted – it will provide a standardised protocol for ACs to control APs (configuration, authentication, session key distribution etc.) and will, eventually, allow interoperability of APs and ACs from different manufacturers. Figure 3.29 shows the overall architecture. Note that in the IETF terminology APs are now called Wireless Termination Points (WTP).

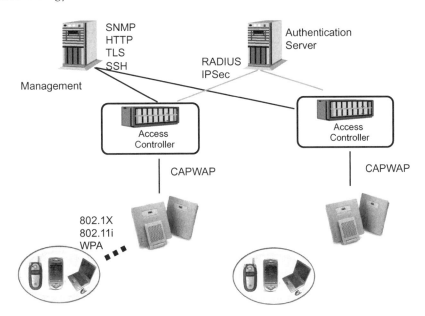

Figure 3-29. CAPWAP – the IETF protocol from Access Point Controllers to APs. (Source: Author. Reproduced by permission of © Dave Wisely, British Telecommunications plc.)

It is also worth noting that WLAN to WiMAX or 3G handover will probably be just as important as WLAN to WLAN handover – since WLANs will (short of new spectrum and higher powers) never cover more than a fraction of urban areas. This "vertical" handover has a number of interesting solutions that are dealt with in Chapter 6.

3.12 Battery Life

One of the major problems with WLANs is that they are very battery intensive. On fixed PCs, game stations and cameras it doesn't matter and on laptops it's OK because they need charging regularly (although having the WLANs on continuously on the laptop I am using to type this book shortens battery life by 30%!). It is PDAs and phones that really suffer with high WLAN power consumption – with various reports of standby times as short as 7–10 hours for WLAN mobile phones. Even the BT Fusion phones (that offer dual GSM/WLAN capability) – which have been extensively tweaked, have much shorter standby time than cellular-only phones. WLAN needs more power than Bluetooth and not many people I know turn on Bluetooth permanently on their phones.

The reason for the high power consumption is not hard to discover. If you have a WLAN – equipped phone how do you receive incoming calls? Well the call is probably controlled by SIP (see Chapter 6) and you have a SIP client running in the background listening on the well known TCP port for SIP messages but that SIP message is wrapped up in TCP and then IP and when it reaches the access point the SIP/TCP/IP packet is placed in a MAC frame and sent out but – and here is the rub – in order to determine if the frame is for your station it has to be received, decoded and the MAC address read. This applies to every single frame transmitted whilst the terminal is associated and authenticated with the access point – even if the terminal doesn't receive or send any packet for (say) 24 hours it will still have to look at every frame. The fundamental difference with a cellular system is that WLANs have two modes – on (associated and authenticated) and off (i.e. switched off) whereas cellular systems have three – on (ready to transmit/receiver), off and idle. The idle mode is a special low power mode that allows phones to listen only on certain channels (the paging channel) – when a voice call is signalled to the HLR it knows the location of the phone only to within a paging area (several cells in size – whenever the mobile changes paging area it updates the HLR – but this is very infrequent). A paging request is broadcast to all the cells in the paging area and the mobile responds with its current location (i.e. cell) and moves from the idle state. This idle mode is essential for low power consumption in mobiles.

Not surprisingly there is a move to develop an idle mode/paging system for WLANs – this is the work of the newly formed 802.11v working group (and, before you ask, yes it has reached z !!!). Figure 3.30 shows an early architecture. Things are quite sketchy at this early stage but one of the possible elements of

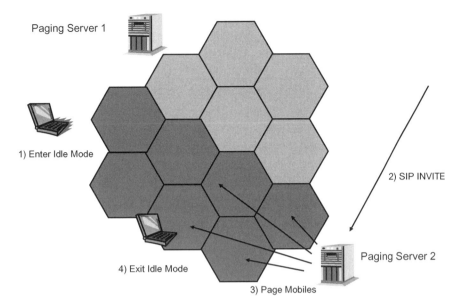

Paging Server 1

1) Enter Idle Mode

2) SIP INVITE

Paging Server 2

4) Exit Idle Mode

3) Page Mobiles

Figure 3-30. 802.11v Idle mode and paging architecture. (Source: Author. Reproduced by permission of © Dave Wisely, British Telecommunications plc.)

802.11v (that my group just happens to be working on) is a small SIP server – either running on an access point or shared between groups of access points. Terminals register with the SIP server and determine a profile of when the SIP server should "wake" them from their low power idle mode. The SIP server would then perform screening – based on time of day, caller identity, remaining terminal battery power etc., before paging the WLAN to wake up. Early prospects suggest a ten times power gain – you can read more about this exciting area of research at Ref. 17.

3.13 Go Faster WLAN – 802.11n

802.11n is the IEEE's group for higher performance WLANs – meaning higher data rates, longer ranges and more reliable communication. The main technical innovation in 802.11n is the introduction of advanced antenna techniques – similar to those described in the chapter on WiMAX (Chapter 5). Multiple transmitters and receivers at the AP and (optionally) at the client can detect and separate multipath reflections that normally cause interference in standard WLANs. In the diversity mode – for example – an 802.11n AP can transmit signals on different paths to a client. Non 802.11n clients simply receiver both signals together. This is just like having two APs transmitting the same information but the diversity gain is that – if the antennas are separated by more than half a wavelength (i.e. more than 6 cm for the ISM band) – then the fading for each signal is totally independent and when one signal is poor (through random fading) the odds are the other has a high signal to noise ratio. The more paths there are the more diversity is achieved and the better the signal to noise ratio. If the terminal is also equipped with multiple transmitters and receivers then it is possible to establish multiple independent communications channels between the AP and terminal (this is explained for WiMAX in Chapter 5) and the overall data rate can be increased by a factor of 2 or 3 in practice. The other major technical innovation that 802.11n introduces is "channel bonding" – which simply means that 802.11n kit can use two independent channels simultaneously (i.e. 40MHz of the ISM band). In fact some proprietary "super g" products are also available that combine channels.

There have been two main proposals in the 802.11n group that have taken a long time to harmonise into a single standard. At the time of writing (Q2 2008) the standard is not due to be ratified until the end of 2008. For the time-being, however, users are able to buy what is labelled "pre-n" kit (such as that in Figure 3.31) including both APs and PCMCIA cards and USB WLAN adaptors.

Figure 3-31. Pre-n router with triple diversity (author). (Source: Author. Reproduced by permission of © Dave Wisely, British Telecommunications plc.)

In our tests across various house types and sizes we have shown pre-n kit offer diversity for non 802.11n clients which results in a larger coverage area with more continuous coverage (i.e. fewer places with poor reception) but not much higher bandwidth. By adding 802.11n cards from the same supplier as the AP we were able to get significant increases (up to twice) in the maximum data rate.

Of course 802.11n is more expensive than 802.11a/b/g needing more complicated radios and more processing. In addition they use (with channel bonding) more of the very scarce spectrum – so they might not give such good performance in areas of heavy interference. You should also be careful to buy equipment from only a single vendor until the WiFi alliance undertakes full interoperability tests – which won't happen before the standard is ratified. For these reasons pre-802.11n tends to be found in homes rather than hot spots or enterprises.

3.14 Interference

WLAN suffer many types of interference from:

- Other WLANs of the same type on the same channel (e.g. both 802.11b).
- Other WLANs of a different type on the same channel (e.g. mixed 802.11g and 802.11b).
- Other WLANs on nearby channels.
- External sources (e.g. Microwave ovens).
- Other users of the (unlicensed) ISM band (e.g. video senders or Bluetooth).
- Licensed users of the ISM band (e.g. railway radio systems).

This is all very different to the careful radio planning of cellular networks. These operate in licensed spectrum where all base-station frequencies are carefully chosen and interference is only generated by frequency re-use in neighbouring cells and this is carefully managed.

Figure 3.32 shows the throughput (for an ftp download) between an access point and a terminal 6 m distant on channel 11 – as an interferer (which is 8 m away and transmitting a small amount of traffic to a remote access point) changes channel. It is seen that the maximum throughput is 5.5 Mbit/s (this is 802.11b) and that putting the interferer on channel 11 (the same channel) results in the two transmissions sharing the bandwidth (even though they are logically separate and use different SSIDs) because the transmissions of one pair are seen as keeping the medium busy to the other pair. This is a crucial point – all stations that are on the same channel and that can detect the headers of received packets – form a single collision avoidance group – as far as the CSMA/CA MAC is concerned. When the channel is changed to 10 the packets from the interferer are not received or decoded by the AP and terminal on channel 11 – as far as the

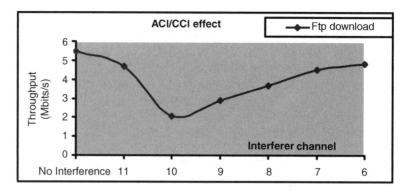

Figure 3-32. Throughput for an ftp session between a terminal and an access point (6 m) on channel 11 with an interferer 8 m away on various channels. (Source: Terry Hodgkinson. Reproduced by permission of © British Telecommunications plc.)

MAC on channel 11 is concerned these frames don't exist and so much of the time the channel 11 packets are sent when there is traffic on channel 10. Because these are so close in frequency the level of lost packets is very high. As the channel of the interferer moves further away from 11 so the level of interference reduces – but note that even at channel 6 there is still some small effect – due to the imperfect action of the band filters (see Figure 3.16).

Worst still is the effect of mixing 802.11b and 801.11g transmissions on the same channel. If you go out and buy a nice new 802.11g system and your neighbour is operating an old 802.11b system on the same channel then you will not get 25 Mbit/s – in fact you might find you get less than 5 Mbits/s. The reason for this seemingly bizarre result is that the two systems form part of a collision avoidance group – i.e. they both detect the frames of the other system and defer to them. When an 802.11g system wins the right to transmit it can send a maximum length packet in about 0.4 ms. When an 802.11b system wins the right to transmit is takes 2 ms to transmit a max length packet at the 11 Mbit/s rate – but if the system is close to its range limit it might only be working at (say) 2 Mbit/s – since the rate is variable depending on the signal to noise. At 2 Mbit/s it will take more like 10 ms for a maximum length packet to be sent. Now, the CSMA/CA algorithm will (when both systems are trying to send data on the same channel) result in equal chances of winning the right to transmit but once that right is won the 802.11g system will use up 0.4 ms but the 802.11b system will use up to 10 ms – so the average throughput for both systems will be about 1 Mbit/s.

Figure 3.33 shows the effect on WLAN throughput of a video sender on WLAN throughput and signal to noise. Video senders – if you have never used one – can relay a TV signal from a video recorder or remote security camera to location 10–30 m away. Typical applications include watching video upstairs when a player is downstairs or monitoring an outbuilding for, the all too common in England at least, yobbos trying to break in. These systems also use the ISM band and the effect – as shown in Figure 3.33 can be devastating to

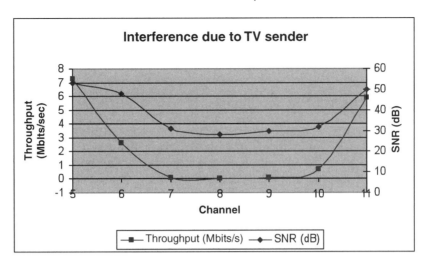

Figure 3-33. Effect of a video sender (centred on channel 8) on WLAN throughput and signal to noise for different WLAN channels. (Source: Terry Hodgkinson. Reproduced by permission of © British Telecommunications plc.)

WLAN throughput with almost five channels with very low or zero throughout. The reason for the low throughput is that the gap between the video frames is shorter than SIFS and the WLAN never sees the channel as free!

The upshot of all this is that you should be quite careful when setting up a new WLAN – clearly you should do a survey of what your neighbours are running – a convenient tool for this is Net Stumbler (Ref. 12). Ideally you need to be three clear channels away from any existing WLANs in the ISM band and well away from any video senders. You could do worse than set up the access point and a lap top for Internet access and time how long "large" www pages (such as e-bay – about 500 k) take to load for all available channels – this should also be compared to plugging directly into the WLAN router with an Ethernet cable – there should be no significant difference if your WLAN is working OK. Don't forget to add security (ideally WPA2) – there are many tools in Windows/Vista to help with this – such as making a file to run on all the client machines and setting them up the same channel and keys.

Interference is also likely to limit Wireless Cities and is getting worse in urban areas – especially in Europe where "dense build" is the norm. In Enterprises it is not quite such a problem – although we have detected 43 WLAN SSIDs at BT Adastral Park and over 20 at BT Centre. Unsurprisingly, both locations have presented us with problems for running experiments or demonstrations. Systems that worked happily in the evening fail at 9am when all the WLANs are turned on. The only real answer is more spectrum – with the 5 GHz unlicensed bands being the most likely candidates in Europe and Japan. Of course in the US, with house densities a fraction of those in Europe and the 802.11a system with 12/13 non-overlapping channels the situation is not so acute.

3.15 Future of WLANs

So what is the future for WLANs? I think it is fair to speculate on three possible paths:

The optimistic scenario in which coverage from WLANs increases dramatically – with wireless cities expanding, a large expansion in public and private hot spots and all new DSL connections adding WLAN routers. In addition WLANs start to appear in most new consumer devices: phones; i-pods; cars; fridges and so on – with a solution to the battery-life problem for mobile phones. In this scenario there is also a change in user behaviour – WLANs are used preferentially to cellular whenever available – especially for VoIP. New WLAN services appear that are not possible on cellular (e.g. streaming video at 800*600 resolution – 2 Mbit/s, say).

The neutral scenario sees WLAN remain much the same as today – with WLAN coverage limited to hot spots in the usual places (airports, hotels, coffee shops etc.) but with no significant continuous coverage outside of the square mile of a city centre and poor user figures for these. In this scenario WLANs are still a niche feature for mobile phones and only used in specific instances (e.g. in range of your home DSL/WLAN router) for devices such as i-Pods. Consumers make effectively all their voice and messaging on their cellular/fixed phone and only use WLAN for mobile data with laptops or enterprise-targeted PDAs. New WLAN data applications do not develop and WLANs are still used for Internet access and VPN Intranet connections.

The obsolescence scenario: In this scenario many of the roles envisaged for WLANs are replaced by other mobile technologies – in particular WiMAX and 3/3.5G. In this scenario cellular data costs fall substantially to be much closer to WLAN data rates today and cellular connectivity starts to be built into laptops and other consumer devices. WLAN survives in its original guise as a wireless Ethernet – to connect computers in the home or office – but hot spots and wireless cities are replaced by WiMAX or 3/3.5 G data coverage.

It is actually, I think, a critical time for WLANs with forces for all these scenarios at work. The best way to try and forecast the future is to look at some of the evidence and try and weigh it up positively.

Firstly, there is no doubt that WLAN coverage is increasing – Table 3.4 shows some very rapid hot spot growth across Europe from 24 000 in 2004 to 42 000 in 2006. Add that to the 20 or so cities and towns in the UK alone that have been rolled out or are in the planning stage and you get some idea of the momentum behind WLAN coverage. Another significant development (Ref. 18) is that BT now offers its broadband customers the option to share their broadband by having their WLAN routers offer two SSIDs – a home (private) one and a public (BT FON) one – in exchange for wider WLAN access. Currently (Q2 2008) users must opt-in to join BT FON – entitling them to use any other BT FON access point as well as 500 min free on Openzone – and 120 000 have so far done so.

	2004	Share (%)	1H06	Share (%)
T-Mobile	5,218	21.1	7,731	18.4
The Cloud	5,215	21.1	7,475	17.8
Orange	1,687	6.8	4,956	11.8
Swisscom Eurospot	1,929	7.8	2,439	5.8
Telefónica	500	2.0	2,008	4.8
BT Openzone	1,327	5.4	1,737	4.1
Telecom Italia	625	2.5	1,100	2.6
Telia Home Run	842	3.4	939	2.2
Portugal Telecom	429	1.7	911	2.2
Eircom	112	0.5	902	2.1
Telenet	200	0.8	886	2.1
KPN	180	0.7	851	2.0
Other	6,475	26.2	10,066	24.0
Total	**24,739**	**100.0**	**42,001**	**100.0**

Table 3-5. WLAN hot spot growth in Europe – Source IDC 2006.

On the other hand there is reason to be cautious about WLAN coverage. Look at Table 3.5 – in 2004 52% of hot spots were owned by nonmobile operators (The Cloud, BT and others) – by 2006 this was down to 45%. This indicates that the industry is starting to consolidate – with fewer small players. It may be analogous to what happened with ISPs – at the height of the dot com boom there were thousands, but by 2007 the UK ISP market was dominated by just four players. Many of these operators are beginning to slow their investment in WLANs because traffic growth – whilst initially rapid – has begun to tail off. I was told by a consultant the other day that one operator put this down to "early geek take up" and then a major slowing when "there were no more geeks to sign up"!

Further evidence for caution:

"The City of London **WiFi** network has only registered 6,000 users in its first month out of a possible 350,000, despite offering the service for free. Such disappointing take-up in the Square Mile raises fears that on-street wireless access will not be commercially viable". Ref. 20

The next most important determinant for WLAN growth, in my view, is devices. WLANs are now commonplace in laptops and PDAs but have yet to make it into mobile phones. In late 2005 a survey by BT expected up to 28 dual-mode Wi-Fi/GSM products to be launched in the following 12 months. This seemed to herald a move by WLANs into the "bread and butter" voice market of cellular mobile. For me this is the absolute crucial step for WLANs – if they are to break out from being just connectivity, from being just for Inter/Intranet access, from being in laptops and PDAs only – then they must be used for voice. Voice use in turn implies that WLANs are in most phones, that the WLAN is on most of the time and the phone is configured for WLAN use wherever possible. Unfortunately, from the optimism of 2005, the truth of WLAN voice today is that mainstream phones don't have WLANs – it is mainly confined to smart phones – and even there WLAN battery consumption is too high to allow them to be used more than intermittently. The only one exception is UMA – Unlicensed Mobile Access (see Chapter 7) – which is essentially a GSM phone which encapsulates and transmits the voice and GSM signalling over WLAN when in range of a home base-station or WLAN hot spot. Users get a cheaper rate on these calls – with the Orange Uniq (Ref. 19) being the most successful with 100 000 + subscribers (Q3 2007).

Even with UMA, however, the success of VoIP over WLAN is qualified. Firstly, only the Orange UMA service in France can be said to be a runaway success – similar services by other operators have been less successful (e.g. BT with Fusion – 40 000 subscribers in 18 months); secondly, this is not so much VoIP as "tunnelled GSM" and offers nothing that GSM doesn't offer; thirdly, the range of handsets has been very restricted and the power consumption much higher than standard GSM handsets. Finally, this is an offering from mobile operators – using WLANs simply to isolate zones of cheap tariff and utilise cheaper (IP) backhaul (or at least the customer is paying for it rather than the MNO). After all GSM networks in Europe are not short of voice capacity – and even if they were 3G adds much more.

So WLAN have not yet been able to break into mobile phones in any meaningful way – they are confined to laptops, PDAs, printers, cameras and games consoles and in every case the business model is the same (more or less) – with WLANs being free (e.g. when used at home or in the enterprise) or charged for as access (as in a hot spot). Other than UMA there is no evidence that WLANs are being integrated into wider convergence offerings.

The major reductions in prices for mobile voice and now mobile data (see Chapter 7) is also of major concern to WLANs. In 2006 the cost of mobile calls in Germany dropped 40%. In the UK the price of mobile data has dropped even more sharply. At the same time mobile data rates are increasing with initiatives such as HSDPA and Femto cells (Chapter 4). If these trends continue then much of the rationale for WLANs – higher speeds and cheaper prices – will be undermined. Add in the appearance of 3G/HSDPA chipsets in some new laptops and suddenly WLANs look less attractive. WiMAX – "described as WLANs on steroids" could be the final step in the demise of WLAN hot spots and Wireless Cities if it undercuts WLAN data prices – especially as most laptops and PDAs will have WiMAX built in and given the vastly improved support for power saving and quality of service (specifically for VoIP) that WiMAX offers compared to WLANs.

In conclusion I am slightly negative on WLANs – I think the neutral scenario – of the three presented at the start of this section – is the most likely. WLANs will continue for home and enterprise data distribution but I think hot spots and Wireless Cities will fall victim to 3G/HSDA and WiMAX in the next three to five years.

References

[1] "Wireless LAN – Mixed Message for Mobile", Global equity research, Lehman Brothers, 4 July 2003.

[2] *Business Week*, 25 April 2005 NEWS: ANALYSIS & COMMENTARY "Intel's WiMax: Like Wi-Fi On Steroids". http://www.businessweek.com/magazine/content/05_17/b3930072_mz011.htm

[3] "Optical wireless: a prognosis". Peter P. Smyth, Philip L. Eardley, Kieran T. Dalton, David R. Wisely, Paul McKee, David Woodin Proceedings of SPIE- The International Society for Optical Engineering. 2601, 212–25, 1995.

[4] WiFi Alliance http://www.wi-fi.org/

[5] Western European WLAN Survey 2006, International Data Corporation (IDC), 12 Sept. 2006.

[6] Total Telecom, Tuesday 24 April 2007.

[7] http://www.luchtzak.be/article4331.html

[8] "City of London gets total WiFi coverage" Total Telecom, Monday 23 April 2007.

[9] http://www.palowireless.com/hiperlan2/

[10] "Cluster Based Channel Allocation for Public WLANs", S. Kawade, V. Abhaya-wardhana, T. Hodgkinson, D. Wiselyin Vehicular Technology Conference, 2007. VTC2007-Spring. IEEE 2007 ieeexplore.ieee.org

[11] "Sharing Your Urban Residential WiFi (UR-WiFi)", S. Kawade, J.W. Van Bloem, V.S. Abhayawardhana, D. Wisely, Vehicular Technology Conference, 2006. VTC 2006-Spring. IEEE, 2006 ieeexplore.ieee.org

[12] http://www.wi-fiplanet.com/tutorials/article.php/3589131

[13] *How Secure Is Your Wireless Network? Safeguarding Your Wi-Fi LAN*, Lee Barken, Prentice Hall, 26 August 2003, ISBN-10:0-13-140206-4.

[14] "WEP key wireless cracking made easy", John Leyden. Wed. 4 April 2007, The Register www.theregister.co.uk/2007/04/04/wireless_code_cracking/

[15] *IP for 3G*, D. R. Wisely *et al.*, John Wiley & Sons Ltd 2002, 0-471-48697-3.

[16] http://msdn2.microsoft.com/en-us/library/aa374474.aspx

[17] http://grouper.ieee.org/groups/802/11/Reports/tgv_update.htm

[18] www.bt.com/btfon

[19] http://www.francetelecom.com/en/financials/journalists/press_releases/CP_old/cp060925.html

[20] "WiFi fails to excite Londoners", *Computing*, Thursday 31 May 2007.

More to Explore

WLAN in general:
- www.javvin.com/protocolWLAN.html
- www.cisco.com (excellent white papers)

802.11:
- www.wifiplanet.com/tutorials
- www.ieee.org

WLAN security:
- www.iss.net/wireless/documents.iss.net/whitepapers/wireless_LAN_security.pdf

Regulation:
- www.aca.gov.au/radcomm/frequency_planning
- www.ofcom.org.uk

WLAN hacking:
- www.netstumbler.com
- www.cs.rice.edu/~astubble/wep/wep_attack.html

WLAN business models:
- "WiFi hotspot service strategies" – Ovum March 2005

802.11 working groups:
- http://grouper.ieee.org/groups/802/11/index.html

Chapter 4: Cellular Evolution

4.1 Introduction

This chapter represents the most obvious view of what fourth generation mobile might be – yet another cellular system evolved from existing 3G systems. Since the launch of 3G in 2001 there have been more Releases of 3GPP standards – we are currently finalising Release 8 – and these have brought with them evolutionary changes to existing 3G systems that have come to be termed 3.5G. HSPA – High Speed Packet Access– is (in theory at least) a software upgrade for the 3G network and a new terminal for the user that drastically increases the data rate and reduces the latency of data connections. In the US CDMA2000 EVDO (EVolution Data Optimized) is following the same path. In the normal scheme of things this would have been followed by a full scale upgrade and a major revision to the air interface in five years time or so (deployment in 2013–2015, say). However, this rather leisurely timescale – dictated by the need to get a return on the huge investments operators have made in 3G networks and licences – has been brought forward by the appearance of a very different cellular mobile standard in the last few years – WiMAX. The final part of the chapter will consider both 3GPP and 3GPP2's responses to WiMAX in terms of wholly new air interfaces (Long Term Evolution and EVDO Rev C respectively).

This chapter is first going to focus on enhancements to UMTS and CDMA2000 – the 3.5G evolutions. There are several good reasons to look at 3.5G in our quest for the mercurial beast that 4G is. Firstly, 3.5G systems are the first cellular technologies actually designed to work exclusively with IP applications and to deliver something approaching the experience of fixed broadband. Secondly, 3.5G systems are here and now and available for

IP for 4G   David Wisely
© 2009 John Wiley & Sons, Ltd.

measurements – many of which are presented in this chapter. This is important as it provides solid evidence on which to judge the "claimed" benefits of more advanced systems – like WiMAX and LTE. Finally 3.5G can get a lot better – in terms of throughput, delay and cost per MByte – with a series of upgrades that are (relatively) inexpensive for operators. Wireless cities, DSL broadband and WiFi hot spots are all in competition with 3.5G for Internet data – if 3.5G improves fast enough it might just take significant market share from these alternatives. Personally I am a fan of HSPA – for £15/month in the UK I have HSPA with a data cap of 3GByte and maximum (burst) speeds of 7.2 Mbit/s.[1] Although this isn't quite the whole story – as this chapter will show – it is very competitive with DSL (in the UK about £15/month plus fixed line rental) – implying the "mobile premium" for data is much lower than for voice.[2]

The chapter starts with a description of HSDPA – High Speed Packet Downlink Access – the UMTS downlink speed enhancing technology which is widely available in Europe and Asia Pacific today. That is followed by HSUPA (High Speed Uplink Packet Access) – the uplink counterpart. HSUPA is nowhere near as effective as HSDPA – but in a world where most traffic is in the downlink this is not much of a barrier.[3] There is a small section on EDGE (Enhanced Data for GSM Evolution) – an enhancement for GPRS – mostly because the first version of the i-Phone – that great symbol of mobile Internet access – doesn't use 3G and has to rely on EDGE. Then there is a section on CDMA2000 data evolution. This is smaller than the HSPA section because: there are three UMTS users for every CDMA2000 user, most of the radio technologies and upgrades used are the same and, finally and most importantly, this book has a page limit and I can't include everything. Finally there is a section on LTE and CDMA Rev. C – the longer term evolution technologies of 3GPP and 3GPP2 – seen as a response to the threat of WiMAX.

4.2 What is Wrong with 3G?

Hold on a minute – Chapter 2 left off with a description of the packet data capabilities of 3G – and, after all, it was not supposed to be about voice and SMS but whizzy new services: video calling, football highlights, interactive games. . . If 3G is as good as the marketing hype then why do we need 3.5G at all?

---

[1] Vodafone – Mobile Broadband. Works well where there is 3G coverage (most towns) when it always goes on to HSPA. Out of 3G coverage or deep indoors it falls back to GPRS which is like dial-up.

[2] The mobile premium is how much more expensive a service is on mobile when compared to fixed. This is often impossible to pin down precisely as both network types go in for call packages and options with free or bundled minutes. However, for voice the premium is about 5x for many typical users.

[3] Even with YouTube and other video sharing sites, P2P, file-sharing and so forth the asymmetry remains quite strong in the downlink. Typical figures for BT's network vary between 3.5 to 1 and 5 to 1.

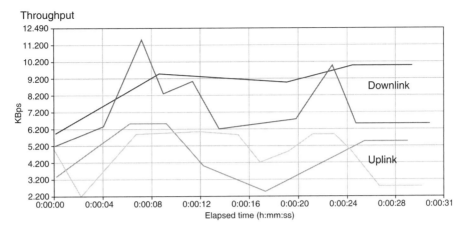

Throughput

Figure 4-1. Typical 3G throughput in London (BT measurements – summer 2007. BT Mobile service). (Source: BT. Reproduced by permission of © British Telecommunications plc.) Note throughput is Kbytes/s – multiply by 8 for Bbit/s.

The first thing that turned out to be more hype than performance from 3G was the data rate. Coverage of 3G networks is still patchy – meaning some of the time data users are back on GPRS rates – and even in major cities real actual throughput is typically in the range 50 kbit/s–300 kbit/s (Ref. 1). Figure 4.1 shows some throughput measurements taken in London in the summer of 2007.

The other side to throughput is latency – the end to end delay from a packet being sent from a game server (say) to it actually reaching your 3G terminal. Again there is quite a lot of variation in 3G systems but values typically lie in the range 200–500 ms. This is a very high latency by IP and Internet standards – meaning that interactive games and real-time voice and video applications are impossible to use on most existing 3G networks. Figure 4.2 shows some typical latency measurements.

The next technical issue with 3G is related to the nature of W-CDMA. In W-CDMA systems the interference caused by any transmission is "smeared" over the entire carrier[4] (see Chapter 2) – with the result that the capacity of the system is essentially limited by the background interference level. Since UMTS employs a frequency reuse factor of 1 (i.e. all the cells/sectors use the same frequency or chunk of spectrum) this interference also comes from all neighbouring cells. However, it is estimated that 70% of CDMA interference is generated within the same cell – one of the reasons a frequency re-use of 1 is possible. In a lightly loaded cell the range of the cell can be much further than when the cell is heavily loaded and the interference level rises – this is known as

---

[4] Each carrier in UMTS is 5MHz. Of course some operators have several carriers but these are totally independent of each other.

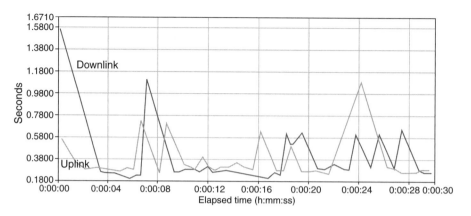

Figure 4-2. Typical 3G latency measurements in London (BT measurements – summer 2007 – these are downlink, one way delays). Source: BT. Reproduced by permission of © British Telecommunications plc.

cell breathing (Figure 4.3). Cell breathing makes network planning more difficult and can reduce data rates very drastically at the edges of cells – often in unpredictable patterns.

In addition to causing cell breathing the fundamentals of W-CDMA means that to achieve a decent capacity the system has to support soft handover: a terminal is receiving the same radio frame from a number of base-stations and combining them to achieve a much higher signal to interference ratio than would be the case for any one of them. This allows the system to run at higher interference levels (i.e. higher capacity) but demands that voice and data streams are delivered to and from base-stations with very precise timing. In 3G

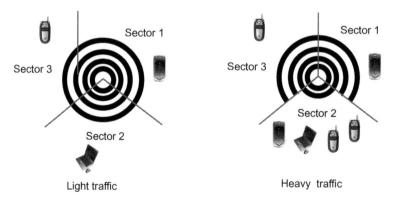

Figure 4-3. CDMA cell breathing. (Source: Author. Reproduced by permission of © Dave Wisely, British Telecommunications plc.)*

- **Korea 39%** 15 M EV-DO/WCDMA out of 38 M total;
- **Japan 34%** 29 M WCDMA/EV-DO out of 85 M total;
- **Italy 16%** 12 M WCDMA out of 70 M total;
- **Portugal 11%** 1 M WCDMA out of 10 M total;
- **UK 8%** 5 M WCDMA out of 65 M total;
- **Austria 8%** 690 K WCDMA out of 8 M total;
- **Sweden 8%** 830 K WCDMA out of 9 M total;
- **Australia 6%** 1 M WCDMA out of 17 M total;
- **Hong Kong 5%** 400 K WCDMA out of 8 M total;
- **USA 3%** 5 M EV-DO out of 190 M total;

Table 4-1. 3G take up in 2006 – Sources: Informa, EMC Database, Global Mobile.

this is solved by using ATM switching – ATM being able to control the delivery of frames very tightly in time. In hindsight, however, ATM has been declining (replaced with Ethernet or MPLS) and the ATM legacy of 3G is seen as inflexible and costly going forward.

In addition to network issues there have also been terminal problems – with the industry taking five years or so to come up with handsets that are able to match GSM for weight and talk/standby performance. IPR (Intellectual Property Rights) have also been a major cost of 3G – with an estimated 8–28% (Ref. 2) of each handset going in licensing fees.

3G take up has been strong in the Far East and much slower in Europe (Table 4.1). Some commentators (Ref. 3) are of the view that Europe and the US will simply follow the take up in the Far East – lagging two to three years behind (Figure 4.4). I am slightly more sceptical – conditions in Japan are unique with low broadband penetration, long commuting times and a generally greater enthusiasm for new gadgets and services.

One of the main reasons for the lack of take up of 3G has been the lack of new, let alone "killer", applications. As will be explored in much greater detail in the chapter on services (Chapter 7), 3G is still used predominantly for voice and

	Months from Launch	3G Penetration
Sweden	28	11%
Portugal	30	18%
Austria	32	15%
UK	36	17%
Italy	36	22%
Japan	54	38%
Korea	52	44%

Figure 4-4.   Take up rate against years since launch for 3G (Figures from Tomi Ahonen). (Source: Author. Reproduced by permission of © Dave Wisely, British Telecommunications plc.)

messaging. I conducted a poll of 100 researchers working at BT and, of these, 37 had 3G phones but only five of them said they used them for anything other than voice or messaging!

One of the reasons for the lack of data application has been cost – with per MByte prices hundreds or thousands of times higher than for DSL, bandwidths ten times slower and latency ten times longer, it is no wonder mobile users have shunned the Mobile Internet. This has changed recently, however, with the introduction of Mobile Broadband (plug-in dongles for laptops) that offer much lower per MByte charges. Mobile Broadband will be covered later – but, the fact that cheap data has been offered only several years after the launch of 3G remains one of the reasons why 3G data has been slow to take-off.

3.5G and LTE are the mobile industry's answers to fixing these issues – WiMAX is the computer industry's answer – in the next section and chapter, respectively, you can read about how they work.

4.3 HSPA

HSPA (High Speed Packet Access) is a series of incremental changes (upgrades) to the UMTS mobile network – relatively minor in comparison to the overall build and running costs of 3G – as well as new terminals for users. HSPA is only about enhancing the packet switched element of 3G – the circuit element is unchanged – with higher overall capacity (i.e. better spectrum efficiency), higher data rates for users and lower latencies. In this section we will first look at HSDPA – High Speed Packet Downlink Access – as this is the first to be deployed and is the area where most efficiency improvements can be made. HSUPA – High Speed Packet Uplink Access – with improved uplink data rates and latencies will lag one or two years behind and offer less impressive improvements. Given the asymmetric nature of most existing mobile data applications (e.g. music or web page downloads) this is not seen as an issue. The industry is geared up for most new terminals to be HSPA-enabled by 2009 (Figure 4.5). In addition there is gathering momentum for HSPA+ : further enhancements to HSPA with multiple transmitters and receivers as well as new coding schemes – we will also look at the prospects for this at the end of this section.

4.3.1 HSDPA Basic Principles

HSDPA introduces four new technical innovations over the packet switched radio network of R99 – this involves upgrades at the Node Bs, RNCs and new user terminals. For some networks this has been effectively only a "firmware" upgrade. The antenna, backhaul, circuit switched radio network and the whole core network is unchanged by these upgrades.

The first innovation of HSDPA is the use of a single high speed data channel that is shared between all the mobiles– in contrast to R99 in which each

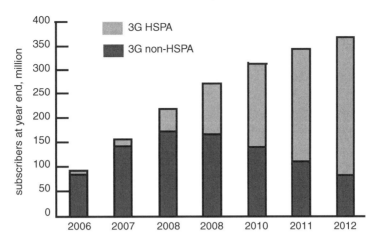

Figure 4-5. Growth of HSPA-enabled handsets (courtesy of BTTJ – source UMTS forum report 19). (Source: BTTJ. Reproduced by permission of © British Telecommunications plc.)

terminal gets its own dedicated data channel. This can be inefficient because the propagation characteristics of radio channels can change rapidly – especially for moving users. The characteristics of the transmission from the base-station to the mobile (data rate, modulation, error coding etc.) is set as a compromise that is often not optimum for the precise radio conditions at any given instance. In HSDPA the capacity from the individual channels is pooled and it is scheduled to be used by different terminals at different times depending on the precise radio propagation they have with the base-station. Clearly the idea is to allow terminals with good signal to interference ratios (S/I)[5] to receive data at the highest possible rate, whilst those with poor signals do not have data sent by the base-station until the S/I ratio improves. Figure 4.6 shows two users who are moving and whose S/I varies. By scheduling them only when the S/I is high it is possible to double the overall throughput compared to the R99 arrangement of separate channels. In practice this works better with a larger number of mobiles transmitting – where it is always possible to find a terminal with excellent S/I – and for situations where the terminals are moving and sampling a range of radio conditions. The overall capacity gain – known as the user diversity gain – is reckoned to be 30% to 100% over 3G (Ref. 4).

To make user diversity work successfully, however, the scheduling interval must be reduced to 2 ms – this is very important as in R99 the Transmission Time Interval (TTI) as it is known is: 10, 20, 40 or 80 ms. During these times the radio conditions of moving terminals can change rapidly and in R99 compromise radio settings are used that do not exploit windows of very high S/I. In addition

---

[5] Strictly signal to interference plus noise – but in CDMA systems the dominant contribution is same-cell interference.

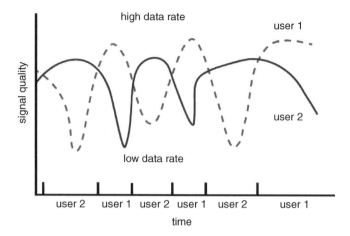

Figure 4-6.   User diversity in HSDPA (courtesy of BTTJ). (Source: BTTJ. Reproduced by permission of © British Telecommunications plc.)

R99 sets up a dedicated channel for each user – even for packet traffic. This dedicated channel is a compromise (of modulation and coding and, hence data rate) to fit the average conditions that the mobile experiences. That means that periods of good propagation are never exploited and periods of very poor propagation generate many errors and contribute to delays. To achieve scheduling at 2 ms interval the scheduling is moved from the RNC to the Node B and a new QoS provisioning function is also required.

The second innovation is to introduce higher order modulation schemes and new coding ratios. These have the effect of lifting the maximum achievable data rate as well as offering a much finer gradation of possible combinations of coding/modulation. 16-QAM (Quadrature Amplitude Modulation) is added to QPSK (Quadrature Phase Shift Keying) with code rates from $1/4$ to near 1. Since CDMA systems operate power control (for reasons outlined in Chapter 2) – it might be thought that these new combinations were self-defeating – meaning that to use them requires more transmitted power than necessary and so will generate more interference and so reduce overall capacity. If power control were perfect then it is true that you might well end up with one high rate channel generating the same interference as four low data rate channels! However, power control is far from perfect. In the uplink a dynamic range of 71 dB is possible but for the downlink the dynamic range is limited to 10–15 dB.[6] For

----

[6] In the downlink direction you might think that since each code is received by only one mobile the code power could be just sufficient to be decoded above the noise floor of the receiver. This would be true if the codes were really orthogonal but they are not quite (due to multipath, amongst other causes). If your receiver is very close to the base-station you might easily detect a low signal level if you were the only user. If, however, there are 10 users far from the base-station they will need very high transmit powers and, because the codes are not orthogonal, you will experience heavy interference. This is why the downlink dynamic range is so limited and is called the "near-far" problem.

users close to the base-station the transmit power is much higher than needed for R99 maximum data rates – so the introduction of higher rates means that more data can be transmitted for the same amount of interference: in other words the capacity can be increased. It is possible for terminals to have a raw data rate of up to 14 Mbit/s – although before you get all excited about HDTV on your mobile!! – this is likely to apply only during one TTI – but the point is a terminal can receive a lot of data in 2 ms at 14 Mbit/s – 140 Kbyte (if you assume overheads are 50%) – larger than most web pages.

The third innovation that HSDPA brings is fast link adaptation which varies the amount of error coding on a given channel depending on the channel conditions. With a much wider choice of modulation and error coding – over a much shorter interval – HSDPA can exploit favourable radio conditions as well as getting a higher throughput in less favourable ones by better matching the modulation/error coding in use to the actual conditions. Figure 4.7 shows link adaptation for a single user in varying radio conditions. (Remember this is a shared channel and, normally one user might get only intermittent use but here we are imagining a single user getting the whole channel top show the operation of link adaptation).

Finally, HSDPA introduces something called a fast hybrid automatic repeat request (HARQ). This is related to what happens when a frame is lost because – despite the error coding – it can't be recovered due to bit errors. In R99 radio

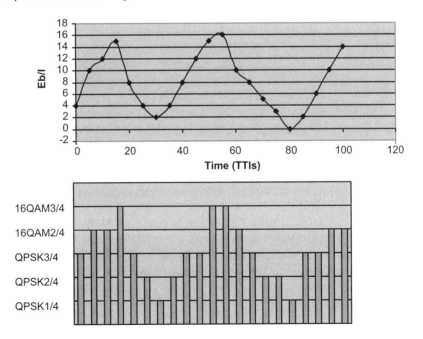

Figure 4-7. Fast link adaptation during transmission to a single mobile user over a prolonged period – showing how the modulation/coding is adapted by the BS – using measurements from the terminal every TTI. (Source: Author. Reproduced by permission of © Dave Wisely, British Telecommunications plc.)

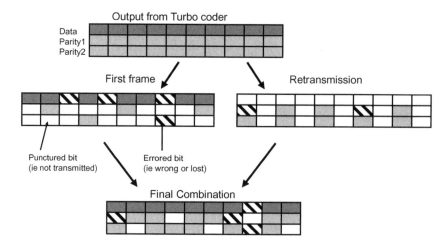

Figure 4-8. HARQ with non-identical re-transmissions – incremental redundancy.
(*Source*: Author. Reproduced by permission of © Dave Wisely, British Telecommunications plc.)

frames that can't be recovered error free are discarded at the physical layer of
the terminal. Recall (Chapter 2) that R99 data can be run in a number of modes
at the RLC (Radio Link Control Layer) that runs between the terminal and the
RNC including unacknowledged (no ACK from the terminal to the RNC) and
acknowledged (an ACK from the terminal to the RNC that frames have been
successfully received). In acknowledged mode the additional latency generated can be quite substantial since the RNC (the Radio Network Controller)
which is typically located in a major city and remote from the base-station site.
For an IP packet – fragmented over several radio frames – the delays involved in
the detection and re-transmission of the constituent radio frames is a major
cause of the high latency figures for 3G (200–400ms one way). In HSDPA this is
addressed by adding a fast hybrid automatic repeat request to the Node B (the
fast part) and by retaining any errored frames in memory within the terminal and
combining it with one or more repeated frames which might also be errored (the
hybrid part).

   As an example Figure 4.8 shows two frames – with the three error coding bit
types produced by the turbo coder (HSDPA uses only turbo coding – another
advantage over the convolutional codes of R99[7]). Two frames – both beyond
recovery are received and stored by the terminal. They are then combined to

---

[7] There is a fundamental limit, called the Shannon limit, to the data rate that can be achieved
over any given communications channel. Not surprisingly the higher the bandwidth of the
channel and the lower the noise/interference, the higher the data rate. However, to actually
achieve data rates approaching the Shannon limit requires very complicated coding schemes
for the data and error correction (e.g. Turbo codes). These perform better than simple (e.g.
convolutional codes) but require more processing and higher power consumption. Again the
march of Moore's Law has permitted these much more complex coding schemes.

produce a single – error-free – frame. If, after a number of retransmissions, the HARQ process fails then the RLC acknowledged mode will cause a re-transmission from the RNC. HARQ is the major feature responsible for lowering the latency in HSDPA.

4.3.2 Channels in HSDPA

Channels are how mobile systems signal, control, allocate and transport bits across the air interface. They support functions like broadcasting and paging and are used to allocate voice and data bandwidth. Sounds simple? However, in UMTS the channel structure is quite confusing. For a start there are: logical, transport and physical channels (Figure 4.9). Logical channels are grouped by the type of information transmitted – they are mapped by the transport channels onto the actual physical channels that are used to convey the information to and from the mobile. Each of the physical channels has its own characteristics – maybe broadcast, fixed coding rate etc. – that is suited to its purpose. The easiest way to understand these is by looking at the example of the downlink – from base-station to mobile. The key logical functions the network needs to perform are:

- Informing the mobile about the environment – such as allowed power levels and the scrambling codes of neighbouring cells/sectors. This is the Broadcast Control Channel (BCCH).

- Paging the mobile when an incoming call/session is received. This is the function of the Paging Control Channel (PCCH).

- Performing control functions common to all mobiles. These are carried out on the Common Control Channel (CCCH).

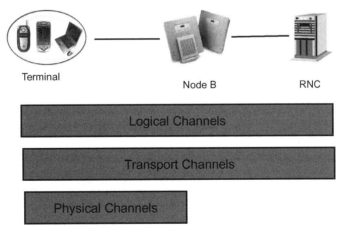

Figure 4-9. UMTS channel architecture. (Source: Author. Reproduced by permission of © Dave Wisely, British Telecommunications plc.)

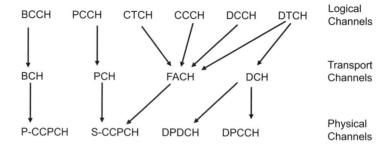

Figure 4-10. UMTS downlink channel structure. (Source: Author. Reproduced by permission of © Dave Wisely, British Telecommunications plc.)

- Setting up voice and date connections. In R99 these are always dedicated connections (i.e. one per mobile per session) and are controlled by the Dedicated Control Channel (DCCH).

- Carrying traffic. Actual user traffic is carried on the Dedicated Traffic Channel (DTCH).

- Carrying traffic to a group of mobiles within a cell – this is the role of the Common Traffic Channel (CTCH).

At the transport channel level these functions are mapped onto four transport channels:[8]

- Broadcast Channel (BCH);

- Paging Channel (PCH);

- Forward Access Channel (FACH);

- Dedicated Channel (DCH).

Figure 4.10 shows the relationship between these channels and the under-lying physical channels. Readers interested in the nitty-gritty detail of these can see Ref. 5 or look back to the more specific description at the end of Chapter 2 – particularly in relation to the physical channels. The broadcast and paging channels are fairly obvious and don't change in HSPA. The FACH is used for transferring very small amounts of data, for first setting up a connection and when the cell is idle and has no other transport control channel available. The key, however, to understanding the improvements of HSDPA is the operation of the DCH. Each mobile receiving downlink data, voice or video, is allocated a DCH. The DCH is carried on the DPDCH (Dedicated Physical Data Channel) and DPCCH (Dedicated Physical Control Channel). The DCH dedicates

---

[8] There was also a Downlink Shared Channel (DSCH) in the R99 specifications but this was never implemented and has been replaced in R5 by a new high speed shared channel of HSDPA.

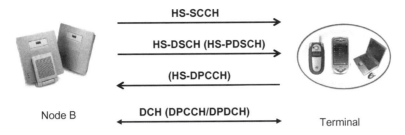

Figure 4-11. Transport Channel structure in HSDPA. (*Source:* Author. Reproduced by permission of © Dave Wisely, British Telecommunications plc.)

resources to a mobile for a fixed time interval, uses a fixed spreading factor and is based on the peak data rate expected. It is scheduled from the RNC and deals with errors at layer 2 only. The maximum data rate so far implemented on the DCH is 384 kbit/s.

Figure 4.11 shows the transport channels involved in HSDPA operation. In Release 5 (R5) HSDPA the DCH is still required to carry the Signalling Radio Bearer as well as any voice traffic. The DCH is also needed for uplink user data.[9]

The most important channel is clearly the HS-DSCH (High Speed Downlink Shared Channel). This runs between the Node B and the terminal – as opposed to the RNC and the terminal for the DCH. The Node B estimates the propagation characteristics for every terminal – based on uplink feedback – and then schedules transmissions, based on a scheduling algorithm to each terminal in turn with a suitable modulation/coding combination. The key question is how does an individual terminal know when – and with what modulation – a transmission is aimed at it? This is the role of the HS-SCCH (High Speed Shared Control Channel) – sent with a fixed spreading factor and QPSK modulation for robust reception in all conditions. The HS-SCCH carries the information needed for the terminals to determine their slots (TTIs) and modulation/coding used.

Finally, in R5, we have the uplink HS-DPCCH (High Speed uplink Dedicated Physical Control Channel) – this allows terminals to feedback ACK/NACKs from the HARQ operation and to report their current propagation characteristics (essentially the S/I ratio). This operation could have been carried on the DCH but that would have meant that soft handover would have had to have been supported. Instead fast cell reselection is supported by this scheme and the R99 DCH left unchanged. In R6 this channel is improved by avoiding periods of having to signal "no transmission" in the ACK/NACK slot when frames are not sent – which was the case in R5. This reduces the peak power required and the need for the Node B to distinguish between three states (No transmission, ACK and NACK).

---

[9] R6 does away with these restrictions – paving the way for VoIP – with a fractional DCH (F-DCH) for the signalling and HSUPA for the uplink packet traffic.

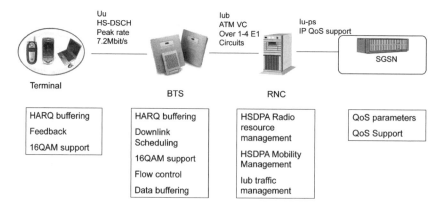

Figure 4-12. Overall HSDPA architecture. (Source: Author. Reproduced by permission of © Dave Wisely, British Telecommunications plc.)

In R6 new channels are introduced for uplink packet transmission without the DCH – which are detailed below. R6 also brings the F-DPCH (the Fractional Dedicated Physical Channel – that allows many users to share a single spreading code) – it was considered that large numbers of VoIP users could not be efficiently supported on the R99 DPCH and so a "cut-down" version – with only power control – has been added.

Figure 4.12 shows the overall HSDPA architecture – with the various functions added.

4.3.3 Handover in HSDPA

HSDPA does not support soft handover. In R99 the dedicated data channel (DCH) can be received simultaneously from up to six Node Bs within the active set. In HSDPA the shared channel (HS-DSCH) is received only from a single serving cell. When a change of serving cell is determined at the RNC (from the received power levels of neighbouring cells reported from the terminal) the terminals and old RNC flush their buffers and the terminal listens to the new Node B. This is a hard handover – break before make – and results in packets being lost. In the RLC acknowledged mode these are retransmitted – although with delay – and in the unacknowledged mode the packets are effectively dropped. A number of small changes have been made to try to minimise the loss of packets on hard handover in HSDPA.

4.3.4 QoS in HSDPA

QoS is difficult term to define – as discussed in the WLAN chapter (Chapter 3) you might perceive the QoS on your 3G phone to be very poor if you don't get indoor coverage or if it takes half an hour for customer services to answer the phone or even if you can't get your favourite TV show. There are many levels of

QoS and these impinge on some of them but here I want to focus on narrower QoS mechanisms – there are plenty of book and column inches written about call centres and don't get me started on a well-known bank's Internet help service – I still have the dents in my forehead from banging it into a solid object "very hard"!!!

Obviously the most important factor in HSDPA QoS is the relationship of network traffic to network capacity. If, as is true for most HSDPA traffic today, the capacity is much higher than the peak hour loading then the network is over-provisioned and no specific QoS mechanism is needed. Operators (in Europe at least) have been busy building "thin and crispy" (after the pizza) 3G networks – i.e. ones that offer broad coverage but limited capacity. As 3G traffic has been largely confined to voice and SMS to date this hasn't really been a problem (certainly the 3G–2G handover has been a much more significant challenge). As data traffic ramps up operators are slowly starting to infill these macro cells with micro and even nano base-stations. Nano base-stations cover a small office block and, crucially, connect back to the RNC over an IP link (see the discussion on Femto cells later in this chapter) – reducing backhaul costs at the expense of handover. WCDMA is well suited to this kind of flexible network build as it has a frequency re-use factor of 1 (i.e. no frequency planning is needed), operators typically have two to four carriers and base-stations are available in a range of sizes.

The Radio Resource Management (RRM) algorithms of 3G have been changed to take into account the new channels and radio resource scheduling needed for HSPA – such as sharing the high speed download channel. Figure 4.13 shows the new split of functionality between the Node B and RNC: what follows is a brief description of these functions – for the gruesome details readers are referred to Ref. 6– a "solid" read if ever there was one.

The RNC is responsible for allocating power and channelisation codes to the Node B for use with HSDPA. Both of these resources are balanced against the

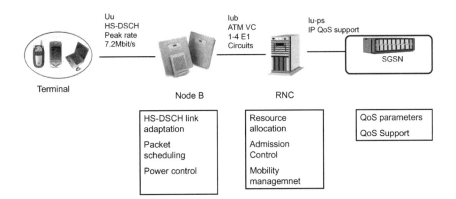

Figure 4-13.   RRM in HSDPA. (Source: Author. Reproduced by permission of © Dave Wisely, British Telecommunications plc.)

power and codes allocated to R99 users. In terms of codes the maximum that a terminal (see next section) can utilise is 15 – so typically 5–10 are allocated for the HS-PDSCH (the High Speed – Dedicated Shared Channel). Far more scarce, however, is downlink power. The RNC is responsible for sharing sector power between HSDPA and R99 users. Typically 3 or 7 W – out of a sector total of 20 W might be allocated to HSDPA. This figure is quite dynamic – giving the RNC a lot of flexibility in deciding the split. The RNC is then responsible for ensuring that the R99 real-time (controlled by the RNC admission control), R99 non real-time (RNC packet scheduler) and common channel allocated power is sufficient and, with the HSDPA power, is low enough for system stability.

The RNC is also responsible for admission control – determining if new users/ service requests can be accommodated and whether they will be allocated to HSDPA or R99 packet or real-time QoS algorithms. The actual algorithms for allocating users/services and then, potentially, adjusting the QoS/prioritisation of existing users/services are not defined in 3GPP standards and are proprietary to each vendor. Measurements (the Node B's total R99 and HSDPA power and the received pilot carrier strength at the terminal – which gives an indication of the quality of the link) and user QoS attributes are then used by the RNC algorithm to decide whether or not to admit the request and what changes to existing QoS are needed to accommodate this. As far as the user is concerned the traffic class and bit rate are what is specified to the SGSN – as shown in Table 4.2.

The Node B is responsible for running the HS-DSCH link adaptation algorithm which adjusts the bit rate, modulation scheme and error coding on each 2 ms transmission time interval (TTI). This is done on the basis of the Signal to Interference ratio as reported by each terminal. The actual decision on which user's packet is transmitted next is the responsibility of the packet scheduler. Again this is a proprietary algorithm – often based on well-known scheduling schemes (Round Robin, Maximum Delay, Proportional Fair Queuing and so forth). These are further modified by the user QoS class and other parameters (e.g. user priority) that the operator invokes. The net result is that QoS in HSDPA is somewhat opaque to the user – they are expected to subscribe to services (Internet, IM, video streaming, VoIP etc.) and the operator will then configure and fine tune the QoS behind the scenes to deliver the services with an acceptable QoS most of the time. Compare this with native IP QoS[10] – user terminals are expected to mark packets (e.g. in DiffServ) or negotiate end-to-end QoS directly (e.g. IntServ); the terminal is in charge of QoS and applications interact with a QoS manager within the terminal. The IP model has failed,[11]

---

[10] Native IP QoS – meaning the IETF "open architecture" approach to QoS – see *IP for 3G* – Wisely *et al.*, Wiley, 2002.

[11] Not quite true – as QoS in IP networks does exist. BT uses DiffServ markings for downstream traffic for special services that the user has paid for – such as IPTV and VoIP. BT also has an extensive MPLS network which creates many QoS classes for core network traffic. However, the point here is that, to date, almost no QoS marking takes place in the end terminal or in real-time (all the MPLS QoS and the DiffServ QoS marking referred to is static – pre-configured in advance).

Traffic class	Conversational class conversational real time	Streaming class streaming real time	Interactive class interactive best effort	Background background best effort
Delay	$\ll$1 sec	<10 sec	approx 1 sec	>10 sec
Example: error tolerant	Conversational voice and video	Streaming audio and video	Voice messaging	fax
Example: error intolerant	Telnet, interactive games	FTP, still image, paging	eCommerce, www browsing	e-mail arrival notification
Fundamental characteristics	Preserve time (variation) between information entities of the stream Conversational pattern (stringent and low delay)	Preserve time (variation) between information entities of the stream	Request response pattern Preserve payload content	Destination is not expecting the data within a certain time Preserve payload content

Table 4-2. QoS classes in HSDPA.

Category	Min TTI	Max codes	Max Data rate Mbit/s
1/2	3	5	1.2
3/4	2	5	1.8
5/6	1	5	3.6
7/8	1	10	7.2
9	1	15	10.2
10	1	15	14.4

Table 4-3. HSDPA terminal capabilities.

however, to gain any traction to date – partly because it is complex and partly because, for it to work, it needs an unholy alliance of: application writers, OS developers, ISPs and interconnect operators to deliver it. I think HSDPA QoS has more chance of commercial success – all of these are at least under the operator's control.

4.3.5 Terminals, Capabilities and Enhancements

HSDPA terminals can be categorised by a number of parameters:

- Maximum number of parallel codes (5 to 15);

- Minimum inter TTI interval (1 to 3);

- Transport channel bits per TTI (7298 to 27952);

- Achievable maximum data rate (0.9 to 14.4 Mbit/s);

- Receiver type (1/2/3 – updated in different releases).

However, this does not mean that users will see application throughputs anywhere near 14 Mbit/s (Table 4.3). This really relates to an instantaneous maximum – a user close to the base-station getting all the available resources for a single TTI. In practice we will see that the type 7/8 receivers in use in early 2008 can give a sustained application level throughput of 2Mbit/s.

The ability to support higher rate terminals and increase the capacity of HSDPA – for a given power allocation – is being enhanced by a series of upgrades. Initially HSDPA has a single Rake receiver[12] (see Chapter 2) in the terminal (called a type 1 terminal) and a two-branch diversity antenna in the base-stations. That is now being upgraded to diversity in the receiver (2 Rake receivers – type 2) and an equaliser (type 3). Diversity works because the CDMA

---

[12] If you are too lazy to look it up a Rake receiver is simply a way of detecting the different reflected/diffracted/refracted version of the transmitted signal by multiplying them with the despreading code delayed by different times. The CDMA codes are carefully chosen to be not only orthogonal to each other but also to themselves when not time aligned.

signal and interference levels vary enough over the separation of the antenna possible in a terminal to make the selection of one or other provide an increase in S/I. In the downlink, in the absence of multipath the different codes are orthogonal. However, multipath destroys this orthogonality and so an equaliser – which restores this – can improve performance.

R6, as previously mentioned, also introduced the Fractional DPCH to allow more VoIP users in a cell and improves the control channel power by changing the NACK/ACK feedback. In addition R6 also introduces support for seamless (low delay, low packet loss) handover for VoIP and reduces the set-up time (i.e. from when an application wants downlink capacity to actually being allocated TTIs) from about 2.3s to 0.8s. R7 goes further and supports more VoIP users in a cell (75 to 100) as well as allowing terminals to stay connected longer – several minutes – if idle. Clearly this is aimed at improving response and set-up times.

Beyond R7 there has been talk, of HSPA+ : to "sweat" the considerable investments in network and licences. HSPA+ is a combination of several technical features:

— Packet call setup time and channel allocation time minimised;

— Enhanced terminals: two-antenna equaliser Rake receiver;

— 2X2 Multiple In Multiple Out (MIMO – see Chapter 5) transmission;

— Higher modulation schemes – including 64QAM;

— A flatter architecture with less protocol overheads.

Headline downlink rates are quoted as 42Mbps (2×2MIMO + 64QAM) and 11.52 Mbps (16QAM) for the uplink. If you want to read about HSDP+ see the Ericsson document RP-060195 from 3GPP. However, there are many caveats to this. Firstly, the range over which 42 Mbit/s would be available is a very small fraction of a typical cell sector. Secondly, terminals will become much more complex and power hungry to service MIMO and high modulations. Thirdly, the cell capacity (as opposed to the headline rate) increase is much more modest for typical cells. HSPA+ might be a "stop gap" to LTE but it is not yet certain that it will be widely deployed. Figure 4.14 shows the likely release dates for the various HSPA upgrades.

4.3.6 HSDPA Data Rates

With the improvements of R5 this is reckoned to give 2–3X the capacity of R99 for packet traffic over a carrier. When the terminals are upgraded to equaliser based antennas and the base-stations employ interference cancellation (i.e. they detect other users' signals and subtract them from the noise) then this will double the capacity of HSDPA again (as shown in Figure 4.15). The changes of R6 will bring a further 20% increase in capacity. The spectral efficiency is then reckoned to be close to 1bit/Hz/cell – so an operator with 10Mz of spectrum

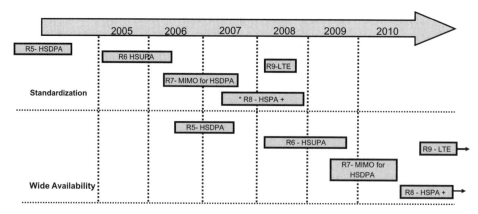

- R5: High-speed Downlink Packet Access (HSDPA)
- R6: High-speed Uplink Packet Access (HSUPA)
- R7: Add Multiple-Input-Multiple-Output (MIMO) to HSDPA
- R8: Evolution HSPA (HSPA+)
- R9: Long Term Evolution (LTE)

Figure 4-14.   3GPP release dates (planned). (Source: Author. Reproduced by permission of © Dave Wisely, British Telecommunications plc.)

(two 5MHz carriers) could expect something like 10Mbit/s continuously (or 5 Mbit/s real application level throughput) – in a sector. When you consider that DSL is often rated at only 1Mbit/s and has a contention ratio of 50:1 then an HSDPA carrier might serve, say, 200 users (see Ref. 7) in a single sector – you can see that these sort of data rates are challenging even fixed line broadband – a topic taken up further in the final section of this chapter.

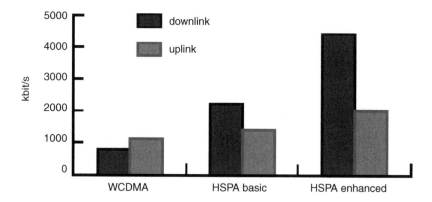

Figure 4-15.   HSDPA cell capacity compared to R99 (courtesy of BTTJ) – one carrier in one sector. (Source: BTTJ. Reproduced by permission of © British Telecommunications plc.)

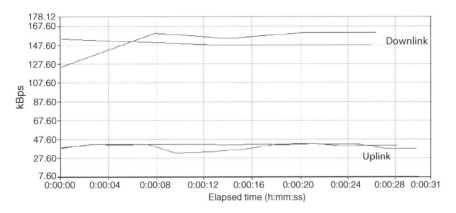

Figure 4-16.   HSDPA data rates – BT measurement (BT Mobile) – summer 2007 with 7.2Mbit/s burst receiver. Not throughput is in KByte/s – for Kbit/s multiply by 8.[13] (*Source:* BT. Reproduced by permission of © British Telecommunications plc.)

HSDPA data rates are quite difficult to determine and, in my view, most figures bandied about should be taken with a pinch of salt. Just as 3G had headline rates of 2Mbit/s but today only delivers 300kbit/s to real applications in real networks – so users should be cautious about promises of 5Mbit/s or more. In terms of maximum user throughput Figures 4.16 and 4.17 show some measurements made in London in the summer of 2007 (it rained a lot so there was not much else to do!). The maximum downlink real application throughput is a sustained 1.2Mbit/s downlink and about 250Kbit/s uplink which I find quite impressive. This certainly implies a cell capacity of about 2Mbit/s (taking all the overheads and headers into account). It is likely that rates will double as new receivers are introduced – offering sustained application rates of up to 2.5 or even 3Mbit/s in the next two to three years.

Rates at the edge of a cell will always be lower than the figures given above. Table 4.4 shows a comparison of HSDPA, R99 and GPRS measured at a rural site (also during the wet summer of 2007).[14]

HSDPA has been widely deployed with 234 HSDPA network commitments in 96 countries and 198 commercial HSDPA networks in 86 countries (Ref. 8). This means that over 90% of commercial UMTS operators have committed to HSDPA – mostly because it is a relatively minor network upgrade that offers significant data capacity and speed benefits. By April 2008 there were 637 HSDPA compatible devices launched (laptops, phones, dongles etc.) with an

---

[13] Further measurements in 2008 suggest a maximum sustained IP throughput of 2Mbit/s is now possible for terminals with a 7.2Mbit/s rating.

[14] Table 4.6 –the section following on LTE also contains further modelling of HSPA rates as the new receivers and version are rolled out.

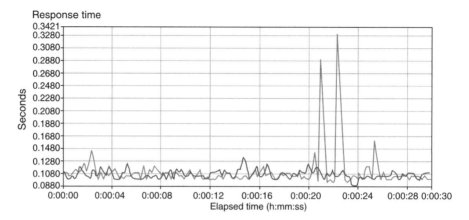

Figure 4-17. HSDPA latency measurement – BT measurement (BT Mobile) – summer 2007 with 7.2Mbit/s burst receiver. Typical latency is 100ms (one way). (Source: BT. Reproduced by permission of © British Telecommunications plc.)

annual growth rate of 150% (Ref. 9) – showing the kind of growth WLAN did a few years ago.

4.4 HSUPA

HSUPA – High Speed Uplink Packet Access is the "little brother" of HSDPA. HSUPA uses some of the same techniques – notably moving fast scheduling to the Node B and a Hybrid ARQ scheme. However, it was felt technically too difficult to create a shared uplink channel. In the downlink the base-station is one transmitter sending data to different mobiles every 2ms or so. This is easy with one transmitter but, in the uplink, would involve coordinating the transmissions of many terminals – terminals that are rapidly moving. WiMAX, as we shall see, uses a very complex scheme to achieve this but this was felt to add too much complexity and cost for HSUPA. In UMTS the uplink transmissions (codes) are not orthogonal and there is not precise time alignment between terminals and the BS – in WiMAX this is essential. HSUPA also does not use higher order modulations – simply because the uplink power control

	HSDPA	R99	GPRS
Max downlink data rate	184 kbit/s	64 kbit/s	31 kbit/s
Max uplink data rate	148 kbit/s	41 kbit/s	19 kbit/s
Average latency (one way delay)	75 ms	200–300 ms	400–1000 ms

Table 4-4. Performance of HSDPA, R99 and GPRS at rural UK site in summer 2007.

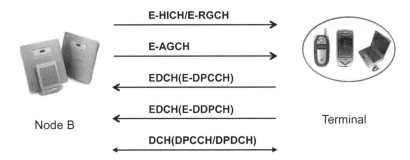

Figure 4-18. Transport HSUPA channel structure. (Source: Author. Reproduced by permission of © Dave Wisely, British Telecommunications plc.)

scheme has a much larger dynamic range (70 dB) than the downlink (15dB). If an uplink has such a good signal to interference ratio that it can support a higher order modulation then it can almost certainly reduce transmit power. This is not the case in the downlink – where the downlink transmit power is much less flexible (15 dB) and the "extra power" would have gone to waste if not used for higher order modulation. HSUPA also adds the ability to shorten the transmit time (TTI) from 10 ms to 2 ms – in simulations, however (Ref. 10) there is little advantage to the shorter TTI until uplink speeds reach 2 Mbit/s and it is likely this will not be the case for some considerable time.

There is – aside from the fast scheduling and HARQ – another important aspect of HSUPA – a complete new transport channel to carry the packet data. Figure 4.18 shows the (complicated) channel structure needed for HSUPA operation when the downlink is on R99. The actual transport channel is the E-DCH (Enhanced Dedicated Channel[15]) – there is only one E-DCH and the terminal must multiplex all uplink packet services onto this one channel: A DCH can be operated simultaneously with the E-DCH but only at 64kbit/s maximum – so that a voice or low rate video call can still continue over the circuit switched path. The E-DCH is mapped to one (or more) physical channels (E-DPDCH – Enhanced Dedicated Physical Data Channels). A new uplink control channel – the E-DCCH (E-DCH Dedicated Control Channel) – is needed to inform the base-station of the format (the transport block size and spreading factor). In addition the channel is used to indicate whether the terminal is content with the current data rate or could make use of a higher data rate. The E-DCCH does not provide information for power control or channel estimation – that is still provided by the DCH.

The Hybrid ARQ works in a similar way to that used for HSDPA – with various options for combining re-transmitted frames. A new feedback channel is needed (the E-DCH HARQ Indicator Channel – E-HICH in the diagram) that provides ACKs and NACKS. There is a slight complication for HSUPA in that, unlike

---

[15] It is called "Enhanced" because that was the original term for HSUPA in 3GPP. The MAC level channel is the E-DCH.

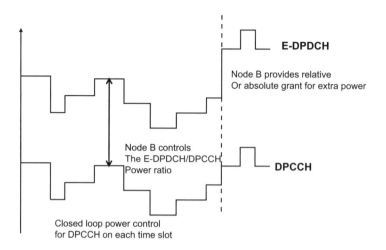

Figure 4-19. *HSUPA power allocation.* (*Source:* Author. Reproduced by permission of © Dave Wisely, British Telecommunications plc.)

HSDPA, it uses soft handover and uplink transmissions from a given terminal are received by multiple base-stations in the active set. All base-stations that are members of the active set can send ACKs – and if a single ACK is received then the frame is not repeated by the terminal. A NACK is only sent by the serving cell.

HSUPA scheduling works by the Node B controlling the maximum power that the terminal can transmit (Figure 4.19). In a CDMA system the total uplink interference must be tightly controlled. Every single transmission contributes to a background level of noise above which each transmission must be "heard". A famous analogy is a party – at first, when there are few guests, the background level of chatter and noise is low and it is easy to have a conversation. Later, when the party is in full swing, it becomes almost impossible to have a conversation – everybody shouts louder but you still can't hear because the background level of noise is rising as fast as your voice. In the same way a CDMA cell has to control the uplink interference level at the base-station. This is, essentially, the transmit powers of all the terminals modified by the channel transmission to the base-station. As the overall power is increased the throughput rises initially but soon users at the cell edge are excluded (called cell breathing – see Figure 4.3) and, eventually, the system becomes unstable because the power control no longer works effectively.

HSUPA shares out this background interference by allocating power to a terminal to use over and above that used on the dedicated physical channel (DPCCH) (Figure 4.19). The closed loop power control keeps the received DPCCH signal at the base-station constant – so this is extra power for the E-DPDCH that represents the terminals share of the extra interference that the Node B has deemed acceptable without causing system instability. The Node B (only one acts as a serving E-DCH cell and controls the power) is able to make relative (step up and step down) and absolute grants of power on the E-RGCH

(E-DCH Relative Grant Channel) and E-AGCH (E-DCH Absolute Grant Channel) channels respectively.

How the Node B divides up the power between the various terminals is (as would be expected) a proprietary algorithm – with information about the user's priority and other policy-based information passing from the HSS (the Home Subscription Server – the 3G user data base) to the Node B. QoS on HSUPA is determined by the terminal's MAC which can select different coding and different power offsets (to improve the bit error rate) of individual TTI frames. Thus delay sensitive traffic would be sent at a lower data rate but with less errors (for a given power) and the terminal will multiplex different flows using a (complex), service-specific, algorithm (e.g. IM might get this QoS but MMS wouldn't). The terminal, however, decides how the power is used and how it is split between applications.

As of Q1 2008 there were 36 HSUPA networks (24 in Europe) in operation around the world (Ref. 11 – Haddon) with 66 devices launched by April 2008.

The following are performance claims for HSUPA (Ref. 12):

- 50–70% improvement in uplink capacity;

- 20–55% reduction in latency;

- 50% increase in uplink data rates.

These will in fact be realised in stages. Table 4.5 shows the terminal types and maximum (burst) rates for HSUPA. In the first deployments terminals with 2 ms TTI will not be available – and the maximum (burst) uplink rate will be limited to 2Mbit/s or so. This translates to a real application throughput of 300–500kbit/s.

As Figure 4.15 shows the cell uplink capacity will approach 2Mbit/s (in 5MHz) – this, in many ways, is more important to operators than headline data rates; allowing many VoIP, WWW and email users (all quite low uplink data rate applications) to share the cell.

The introduction of L1 HARQ is also expected to reduce latency – a L1 retransmission takes about 30 ms compared to 100 ms for L2 retransmission. The 10 ms TTI will limit latency to about 70 ms (one way) but this is expected to fall to about 50 ms with the introduction of 2 ms TTIs.

Terminal category	Max number of codes and spreading factor	Max rate with 10ms TTI Mbit/s	Max rate with 2ms TTI Mbit/s
1	1XSF4	0.72	Not available
2	2SF4	1.45	1.45
	2SF4	1.45	Not available
4	2SF2	2.0	2.91
5	2SF2	2.0	Not available
6	2SF4 +2SF2	2.0	5.76

Table 4-5. HSUPA terminal categories.

The key to the (modest) gains over R99 is really the fast scheduling – this makes better use of the available uplink resources by reacting to the bursty nature of Internet packet data much quicker than the packet transport service of R99 (which is very slow to adapt and inefficient for current Internet traffic).

The question for operators is whether HSUPA is really needed. After all, Internet traffic is highly asymmetric (about 4 to 1 for typical ISPs – why would mobile data services be any different? The current emphasis on mobile services is all on TV and music download (see Chapter 7). Two-way video calling was one of the big flops of early 3G offerings. Video upload from vehicles? Trains and buses with WiFi on board connecting to the Internet are the best ideas I can come up with. However, if mobile operators are going to offer broadband data rates and persuade users to give up a fixed connection completely then a minimum upload of 256kbit/s will be needed and this, realistically, will require HSUPA.

4.5 EDGE

UMTS networks in the UK – six years after launch – still have not achieved anything like the coverage of GSM networks. Even by 2008 current figures are 3 billion GSM/UMTS users with 2.8 billion of these using GSM (Ref. 13). GSM is an effective solution for mobile voice – although CDMA-based solutions provide three to five times as much voice capacity, in a given amount of spectrum, GSM systems have other advantages. Firstly, they tend to have larger amounts of spectrum allocated, the spectrum is at lower frequencies (which means, in coverage-limited areas, that less base-stations are needed) and the handsets are cheaper and perform better (e.g. battery life). As I have said – probably to the point you are getting bored by now? – if all you want to offer in your mobile network is voice and SMS then the answer – by a considerable margin – is a GSM system (at 900MHz if possible). Where a GSM system will let you down is if you want to offer mobile data. GSM has 200KHz carriers which are divided into eight time slots – it is a TDMA/FDD system – and typically each time slot represents one voice conversation. In an urban environment there might be three carriers allocated per sector and three sectors per cell. Each carrier would be reused – but not in a neighbouring cell – typically they are reused every seventh cell (or so) – meaning that each carrier is reused only in 1/20 of the sectors. (15 MHz – a typical allocation in the UK – gives 75 carriers – so there can be three carriers per sector in a reuse of 20 and still have 25 spare). Some networks claim a re-use of only 10 – allowing six carriers per sector. Each carrier can support seven simultaneous voice calls (one time slot is for signalling) and so 6carriers*7calls*3sectors = 126 simultaneous calls per base-station. Given that dense urban base-stations are 100–200m spacing it is easy to see how GSM can easily carry large amounts voice traffic. In the UK there are four GSM operators with 10 000 base-stations each. This is a capacity of 40,000*126 = 5M or so simultaneous calls. Of course this is not quite right as what happens (in the UK) is that a security alert or major transport failure causes a lot of people to make calls from an urban area and a small part of the network is overwhelmed. In Wales there is loads of unused capacity!! But you get the gist – voice is easily handled by GSM. However, it is instructive to compare this with WiFi. Each WiFi access point might handle eight to ten simultaneous voice calls (Chapter 3). To rival the UK GSM network, in pure voice capacity, requires 500 000 access points or so. In fact there are more than this in the UK – 20 000 public ones but 2 000 000 or so private ones. (Ref. 14). However, WiFi coverage remains very patchy and services that offer GSM/WiFi voice solutions (Such as BT's Fusion product – see Chapter 7) have not been a runaway success.

Prior to the introduction of GPRS (General Packet Radio System – see Chapter 2) all data on GSM was handled by having circuit connections open for the length of the data session. GPRS has the advantage, for the operators at least, that users are charged by packet volume – as opposed to time – and that higher data rates were possible (up to about 64Kbit/s). Figure 4.20 shows some measurements made on GPRS that illustrate why it is a frustrating experience

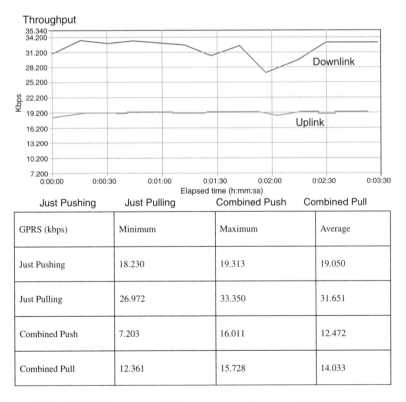

GPRS (kbps)	Minimum	Maximum	Average
Just Pushing	18.230	19.313	19.050
Just Pulling	26.972	33.350	31.651
Combined Push	7.203	16.011	12.472
Combined Pull	12.361	15.728	14.033

Figure 4-20. GPRS data rates (BT measurement summer 2007 (Source: Author. Reproduced by permission of © Dave Wisely, British Telecommunications plc.)

when compared to fixed broadband. Data rates of over 40 kbit/s (at the application level) are very rare – operators can decide how many of the eight time slots in each carrier they will dedicate to GPRS – and this is then shared between users. Each timeslot can deliver about 10kbit/s and most devices are limited to being able to aggregate four slots – meaning 40kbit/s is a typical maximum data rate – something like a dial-up connection – remember those? My Grandma used to tell me about them! The delay is also massive by fixed broadband standards – typically 500ms or longer. This rules out all real-times services (VoIP, two-way video, gaming etc.) and can even be frustrating for web browsing. The original i-Phone only had a GSM radio and is forced to use GPRS when out of range of WiFi – I have tried and swapped my stopwatch for an egg-timer when waiting for maps to load![16]

In Release 99 support was added for EDGE – Enhanced Data rates for GSM Evolution – and this is now widely deployed in the US – so the i-Phone works much better across the pond. EDGE can be deployed in the usual GSM bands (850, 900, 1800 and 1900MHz) and, more or less, doubles capacity for data and triples the maximum data rates. In fact the network is only altered at the BTS and BSC level and, in the case of more modern installations, is a firmware upgrade only. When the network has been upgraded for EDGE it is known as a GERAN (GSM EDGE Radio Access Network). GPRS phones are backwardly compatible with EDGE and so adding EDGE is a fairly minor upgrade. Nevertheless, it is interesting that many European operators have not bothered with EDGE – having paid a King's (and, in some cases, A Queen's, Duke's and Archbishop's as well) ransom for 3G spectrum, they have concentrated on data service launches on 3G. In the US EDGE is much more widespread.

The techniques used in EDGE are, in fact, very similar to those employed in HSPA with three main mechanisms. Firstly, GMSK (Gaussian Mean Shift Keying) – which offers only one bit per symbol – is augmented by 8-PSK (Octagonal Phase Shift Keying) which offers 3 bits per symbol (2*2*2 is 8!). In addition EDGE can select from five coding schemes for 8-PSK and three for GMSK to match the radio conditions between the terminal and the base-station. Again we see that GPRS was conservative – setting a single modulation and coding suitable for cell edges in a coverage limited deployment (meaning cells are widely spaced because they are not fully loaded and terminals are generally far from the base-station). As GSM became successful – beyond its inventors' wildest dreams – it became clear that in urban deployments the system is capacity limited and terminals spend quite a lot of time close to base-stations where the signal to noise/interference level is much better than at the cell edge and will easily support a higher-order modulation. The key to enhanced data rates is the fact that terminals, in real systems, often experience a much better signal to noise/interference level than is needed for GPRS. It has been estimated (Ref. 15) that in a typical urban macro cell, with loading from 15–75% of capacity, that 50% or more have a better signal to noise/interference level than

---

[16] At the time of final writing – July 2008 – Apple had just launched a 3G version.

is needed for GPRS and EDGE offers between two and three times higher data rates than GPRS for these users. With typical rates up to 30 kbit/s per time slot the (absolute max!) rate for EDGE is about 120 kbit/s (at the application level – 200 kbit/s is typically quoted but this overestimates what you will really get for FTP, say), EDGE also introduces the same QoS system as used in UMTS with support for services such as video streaming (VoIP is not possible due to the long delays).

EDGE is not standing still and is set to evolve further. In Release 7 EDGE is enhanced by:

- Adding 16 QAM (Quadrature Amplitude Modulation) which offers 4 bits per symbol – as terminals close to the base-station have signal to noise/interference ratios that can exploit this.

- Allowing reception on more than one carrier – with reception possible on up to 16 time slots and transmission on 8.

- Diversity receivers in the terminal – these improve the signal to noise/interference ratio by mitigating fading. In addition they will be more resistant to interference and allow some carriers to operate in a higher cell reuse pattern – increasing overall system capacity.

- Reducing the Transmission Time Interval (TTI) – as in HSDPA. This will reduce the very long latency of EDGE.

With all these improvements – all relatively trivial in terms of network upgrades – the maximum rate for EDGE gets close to the magic 1 Mbit/s (downlink and 500 kbit/s uplink) speeds that HSPA offer. In Europe and Asia Pacific, however, the emphasis is firmly on 3G.

4.6 EVDO and CDMA2000 Evolution

Chapter 2 describes CDMA2000 and its voice/data networking elements. In essence CDMA2000 is very similar to UMTS – except that it operates with carriers of 1.25 MHz (as opposed to 5 MHz for UMTS). CDMA2000 is widely deployed across the whole of North and South America as the primary 3G technology. It is also deployed alongside UMTS in Asia Pacific (Japan, China and Korea in particular) as well as having deployments in Africa, Eastern Europe and India (see Ref. 16 for the latest list). CDMA2000 has a data capability roughly equivalent to that of UMTS. It also has much the same issues of low maximum data rate (Release 0 of CDMA2000 1X supports data rates of up to 150 kbit/s but averages 60–100 kbit/s in commercial networks), and long latency (500–1000 ms). Figure 4.21 shows the, slightly complicated, evolution path for CDMA2000 networks. There are two paths shown – EVDO (Evolution-Data Optimized or Evolution-Data only) and EVDV (Evolution – Data and Voice). EVDV has been abandoned commercially (hence it is grayed (sic) out)

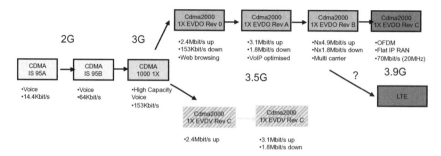

Figure 4-21. CDMA2000 evolution path. (Source: Author. Reproduced by permission of © Dave Wisely, British Telecommunications plc.)

leaving only the EVDO path. The difference with UMTS is that EVDO requires an entire 1.25 MHz carrier to be devoted to it – EVDV would have allowed voice and data to be mixed on the same carrier.

EVDO Rev.0 was first deployed in Korea in 2002 and incorporates much the same techniques as HSDPA and yields a very similar spectral efficiency:

- A shared downlink channel – capable of concentrating resources on terminals with good Carrier to Interference (C/I) ratios.

- Higher order modulations – to exploit situations of high C/I.

- Adaptive modulation and coding to suit the specific radio connection between the mobile and base-station.

- Fast scheduling with shorter TTI – allowing rapid adaptation to changing radio conditions.

- Fast Hybrid ARQ (HARQ).

One of the major disadvantages of EVDO is that the entire carrier (1.25 MHz) is dedicated to data and it is not possible to dynamically allocate resources between data and voice as is possible in HSDPA. Networks quote peak rates of up to 2.4 Mbit/s EVDO Rev. 0 – but measurements (Ref. 17) show typical throughput is about 600 kbit/s downlink and 90kbit/s uplink with a latency of about 420ms round trip. This is roughly comparable to early HSDPA – but the latency is at least twice as high. Figure 4.22 shows the EVDO architecture.

Currently networks such as Sprint and Verizon are busily deploying EVDO Rev. A – this is an upgrade to allow much faster uplink traffic using much the same techniques as HSUPA including:

- Fast Hybrid ARQ;

- Fast uplink rate control and scheduling;

- Adaptive coding and modulation;

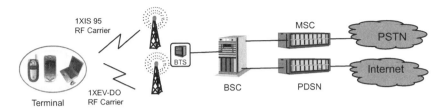

Figure 4-22. CDMA EV-DO Rev 0/A/B. (Source: Author. Reproduced by permission of © Dave Wisely, British Telecommunications plc.)

- Shorter Transmission Time Interval (TTIs);

- Soft handover in the uplink with fast cell selection in the downlink.

It is reckoned (Ref. 18) that Rev. A will increase the uplink burst rate from 150 kbit/s to nearer 1 Mbit/s in real deployments. In addition Rev. A increases the maximum downlink burst rate from 2.45 to 3.1 Mbit/s and substantially reduces latency to (of the order of) 50 ms one-way. More advanced QoS mechanisms allow application-based QoS – meaning that Rev. A should support VoIP and PTT (Push To Talk) easily. Whether operators switch to VoIP is another question – deferred for now until a later chapter.

Moving beyond Rev. A comes. .. you guessed it Rev. B! Rev. B is in fact quite an important step for CDMA2000. The big difference (and a major disadvantage), compared to UMTS, is the small fixed carrier size of 1.25 MHz. Rev. B allows operators to combine up to three carriers for a maximum downlink burst rate of 9.3 Mbit/s. It also offers much greater flexibility in how carriers are allocated to different sectors on the same base-station – with different allocations to different sectors. Rev. B also does away with the need for strict frequency pairing – more capacity can be allocated in the downlink than the uplink if the traffic profile is skewed in that direction with streaming video and music downloads, say. Rev. C is still on the drawing board and uses many of the same ideas as LTE – so I have put them together at the end of the chapter.

4.7 Femtocells

Imagine you are a mobile operator with a 3G mobile network – in most countries this will be a fairly "thin and crispy" (after the pizza) network – meaning that the network is built for coverage and not for capacity (as would be expected in a relatively new and growing network). In fact most 3G networks – certainly in the UK – are under-loaded and have a lot of spare capacity. One of the main reasons for this – as we have seen before is that most of the traffic is voice and SMS. To build on our earlier example – the four GSM networks in the UK were capable of carrying all the voice traffic generated in the country back in 1995. Five shiny new 3G networks – each with 3–5X the capacity of GSM – is clearly overkill for voice and the odd web browsing session. So capacity is not currently an issue but it might be going forward. If mobile data takes off – which it is showing slow signs of doing – then the extra capacity required going forward will be 10–100 times current levels (or even more – in the UK in 2007 it was reported mobile data grew by 800% – albeit from a low base). Currently 16 M broadband users in the UK download, on average, about 1–2 Gbyte per month. How much of this could be off-loaded to the mobile network? Well a typical mobile network in the UK has 6000 3G base-stations – each with three sectors say. At a very maximum they could offer 2 Mbit/s per sector per 5 MHz carrier with HSDPA. So the maximum instantaneous load would be: 6000*3sectors*2Mbit/s/carrier*2carriers = 54 Gbit/s. In practice, however, networks rarely run over 50% capacity and the demand would be uneven, plus the backhaul would need to be upgraded – since this is typically a very expensive part of the cost of running a mobile network. However, let's say that 11Gbit/s (20% efficiency) is possible in reality. How many DSL users would that support? Well typically DSL is 2 Mbit/s with 50 : 1 contention ratio (50 people share 2 Mbit/s) – so 11 Gbit/s is 11 000 Mbit/s/2 Mbit/s*50 = 275,000 users. The need for more capacity is obvious.

Mobile networks are also short of capacity for heavy Internet users. Internet users are currently polarised into light users (who consume about 1 Gbyte/month) do email, book tickets and watch the odd video and heavy users (8 Gbyte or more) who watch videos, download heavily and are P2P (Peer to Peer) file sharers or run servers. Mobile networks just don't have enough capacity to cope with heavy users. Mobile TV to small handsets doesn't require that much bandwidth (the screens are too small) but if mobile networks are to challenge fixed broadband then they will need to cope with much higher data volumes. So it is likely mobile networks will restrict P2P traffic in a way that fixed operators have (to date) not been able to. Another specific issue for 3G is that the 2 GHz frequency band is not good at penetrating buildings to provide indoor coverage – typically signals drop 10 dB (X10) next to a window and 20 dB (X100) inside a house or building – in 3G as the signal level drops so does bandwidth (Figure 4.23).

And then there is the cost! Mobile networks will be competing with WLANs (which are currently expensive at public hot spots but – through initiatives such as FON (see Chapter 3) and lax security – often offer free access). What are the options for increasing data capacity in a mobile network? Remember (Chapter 2) the capacity of a mobile network (simplified) is given by an expression like: No. of users/mile^2 = density of base-stations (no/mile^2) * spectrum (Hz) * efficiency (Bit/s/Hz)/bandwidth of call/session (bit/s)

From this the most obvious answer is to build more base-stations – moving from "thin and crispy" to "deep pan"! That, however, is expensive. New sites, backhaul, protests over masts on schools, vicars wanting their cut for access to the church tower . . . it all adds up. So the next obvious solution is to acquire more spectrum – but that is also expensive and only sporadically available. The next term in the equation is efficiency – so any increase in spectral efficiency

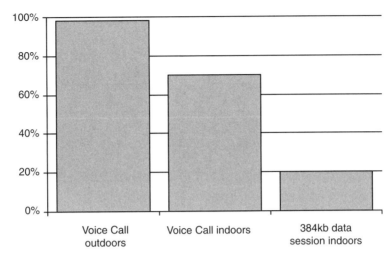

Figure 4-23. Probability of call success for different environments. (Source: BT. Reproduced by permission of © British Telecommunications plc.)

will proportionally raise capacity. HSPA might raise capacity by a factor of 2 in the next two to three years (Figure 4.15) but the current solution for more spectrum and higher efficiency is a brand new air interface called LTE (Long Term Evolution) – which is the subject of the next section but will not be available until about 2011–12. However, coming back to the current discussion – how do I dramatically (say 10 times) increase the capacity of a mobile network without spending shed loads of cash on new base-stations but before LTE arrives? The answer is Femtocells!

A Femtocell is the next progression in the line macro, micro, pico (small base-station often in a hotel or railway station – same as macro/micro only smaller) and nano base-station (very small base-station for an office that connects over IP to the normal mobile network).

A Femtocell is even smaller – covering about 20–30 m – and plugs into a DSL line. As far as the user and the mobile terminals are concerned it still looks just like a normal base-station and all services are exactly the same. There are a number of different usage models for Femtos – but here is a popular one. A user goes to the local electrical store/supermarket and buys a Femto for (say) £60. They then sign up with their mobile operator to have a home Femtocell – for which there is some incentive such as lower charges when connected to the Femtocell or new services (e.g. mobile TV) or just much improved indoor coverage. The mobile operator sends the user a SIM card that is inserted into the Femto to allow the mobile network to control it and prevent rogue base-stations being attached. The user then simply plugs the Femto into their DSL router (it might attach via WLAN) and starts to use it. For the mobile operator there is no outlay (except for a gateway to aggregate all the home cells before they hit the mobile network – since mobile networks are designed for 10 000 base-stations –

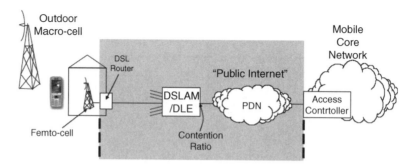

Figure 4-24.   Femtocell architecture. (Source: Author. Reproduced by permission of ©
Dave Wisely, British Telecommunications plc.)

not 1 million) – and no backhaul costs. In theory the capacity of each Femto
could be 2 Mbit/s with HSDPA on a single carrier today and 4Mbit/s in the
future. Figure 4.24 shows the overall set up.

Femtos sound like a great idea for mobile networks but there are a number of
key issues that need to be resolved and are currently the focus of trials and
debate in the mobile community.

Firstly and foremost is the question of cost. Persuading people to go and
spend £50–£100 to buy a device that just improves the mobile network will
need some powerful incentives in lower call charges. On the other hand the
operator might send them out as part of a deal but, again, that cost would need
to be recovered and simply giving away a Femto and getting back less revenue
(because users make the same calls at a lower cost) is not attractive. One of the
main cost elements is the timing of the cell. In a typical mobile handset there is a
crystal oscillator that controls the frequency and timing of transmissions. In
normal operation the mobile frequency is controlled by a much more accurate
(and costly) oscillator in the base-station. This is important as tight frequency
control is needed to allow mobiles to listen to neighbouring base-stations for
handover. The upshot is that you either put an expensive crystal oscillator in the
Femto (£100) or use a cheaper one but accept that handover to neighbouring
cells is not possible. Even at £70 or so the feeling is that the cost is still too high
(compared to £25 or so for a WLAN router).

Secondly, there is the question of attaching the Femto to a mobile network.
They could have an I_{ub} or I_{ur} interface and operate like normal Node Bs but this
is not a scaleable solution as existing RNCs and MSCs etc. are designed to deal
with only a limited number of base-stations. It is generally acknowledged that
some form of gateway or aggregation function is needed.

Thirdly, there is the issue of the ISP backhaul – what happens if the 3G traffic
uses up the monthly limit of the DSL line? Will ISPs be happy to allow mobile
data to transit their IP networks? Then there is the question of QoS – will DSL
lines need to be QoS-enabled to support Femtocells? Early studies suggest
probably not but that does not mean that there will not be issues – especially for
real-time services (voice, video etc.) at peak times. This also goes to the heart of

one of the major cost questions for the mobile network – backhaul. Typically the base-station (Node B or BTS) is stuck out on a farm, up some remote Welsh valley or on a garage roof say. To connect them to a BSC or RNC requires one or more E1 lines. E1 lines are 2 Mbit/s circuit connections – they are dedicated links with tight timing control with very low loss, no jitter and low latency – ideal for the TDM requirements of GSM or the ATM transport of 3G. However, there is only one drawback – they are very expensive – hundreds of times more expensive than DSL. Most "low cost" alternatives to mobile networks for data – and that includes WLAN, Wireless cities and Femtos – are predicated on the fact that this is overkill for most IP traffic. In fact it is now recognised that bursty IP traffic can be statistically multiplexed very successfully on a common pipe and that even voice and video applications – with advanced buffers and heuristic fill-ins for lost packets etc. – work pretty well without dedicated backhaul. So, perhaps, the future of cellular mobile will be decided by backhaul costs? Again there are lots of distorting factors – backhaul in the Welsh valley in order of magnitude more expensive than in London but there seems little chance of introducing regional call pricing. Service pricing – where the price of some bits is higher (e.g. voice) has slipped away from the ISPs and is slipping away from the mobile operators ("three" have put Skype clients on some of their handsets). Finally all-you-can-eat tariffs are unfair to light users – my Dad used to consume less than 1Mbyte a month on broadband. We will continue this discussion in the concluding chapter – but for now note that Femtos are attractive partly because they offload heavy traffic onto a much cheaper form of backhaul.

Finally, there is a regulatory question about Femtos. How can mobile operators ensure that rogue access points aren't used or that malfunctioning Femtos end up blocking mobile signals? Who is responsible for a Femto? If I buy it and plug into my DSL does old granny Stevens – who lives in the flat above – use my Femto as well? If so, what happens if she is making an emergency call and my son knocks the power lead out of the Femto? Who is responsible if another neighbour starts downloading child pornography over my DSL line via the Femto? Does that mean only my own family can connect? But in a block of flats there would be quite a bit of interference if everybody had to have their own Femto? At the moment there are ideas about locking Femtos to networks with SIM cards in them – to ensure they are correctly registered – and to limit access to a set of mobiles (i.e. the family) but trials are taking place to try and resolve some of these issues. Femtos will also have to be location-locked – meaning that they only operate where they are parented to a mobile operator with a licence (so that I don't take my Femto from the UK to Germany say).

There is lots of hype, a few trials and quite a number of conferences on Femtos; there is even a Femto forum (Ref. 19). Whether Femtos actually get deployed in large numbers really depends if people want mobile data in serious volumes. If they do then a whole range of solutions might compete – WLANs, WiMAX and Femtos. If mobile data rates grow only slowly then LTE might introduce enough capacity.

4.8 LTE

LTE – Long Term Evolution – is a 3GPP standards term for a completely new air interface – based on OFDM and smart antennas – and an all-IP network. As we will see in the next chapter these are indeed the key element of WiMAX and, indeed, LTE is seen by some commentators as a response to the emergence of WiMAX which is promoted by a different range of companies and comes from a different standards body (the IEEE). WiMAX is being promoted as a technology that will deliver very high data rates (5 Mbit/s+) to mobile users with low latency with a low cost network. Undoubtedly the choice of WiMAX by Sprint-Nextel as a "4G" technology (Ref. 20) for a $3B roll-out has caused plans within 3GPP to be accelerated. LTE is also motivated by a desire to offer higher data rates than HSPA will be capable of. It will also reduce the Capex and Opex costs of running mobile networks. Much of the drive for LTE is coming from the NGMN (Next Generation Mobile Networks) group which represents an enormous customer base:

- China Mobile Communications Corporation;
- KPN Mobile NV;
- NTT DoCoMo Inc;
- Orange SA;
- Sprint Nextel Corporation;
- T-Mobile International;
- Vodafone Group PLC.

The NGMN group has published a whitepaper (Ref. 21) which summarises the requirements for LTE:

- Significantly increased peak data rates (instantaneous downlink peak data rate of 100 Mbit/s in a 20 MHz downlink spectrum (i.e. 5 bit/s/Hz);
- Increased cell edge bitrates;
- Improved spectrum efficiency (2–4 that of HSPA);
- Improved latency (<30 ms);
- Scaleable bandwidth;
- Reduced CAPEX and OPEX;
- Acceptable system and terminal complexity, cost and power consumption;
- Compatibility with earlier releases and with other systems;
- Optimised for low mobile speed but supporting high mobile speed;

- Capable of flexible operation in different spectrum allocations (operate in 1.25, 1.6, 2.5, 5, 10, 15 and 20 MHz allocation, uplink and downlink, paired and unpaired.

In order to meet these requirements there are a number of candidate schemes being discussed within 3GPP but it is already clear that LTE will use OFDMA (Orthogonal Division Multiple Access – the same as used by WiMAX) in the downlink to achieve the required data rates and efficiencies. It is also clear that the complex soft handover of W-CDMA has been abandoned and that the network will be much simpler and "all-IP". Currently the LTE plan has been endorsed by 3GPP Project Co-ordination Group and an initial study was completed in September 2006. There is great optimism that the relevant standards (Release 8) can be finalised by 2008 and that deployment might take place in 2011–12. Certainly the all-IP network element is lagging far behind. The latest news is that some early LTE systems were demonstrated at the Mobile World Congress (formally 3GSM) in Barcelona February 2008.

In this section we will briefly review the technical innovations that are being proposed for the air interface, look at the all-IP network evolution and, finally evaluate the more commercial aspects of LTE. Also I have included a short section on EVDO Rev C – for adherents of the CDMA2000 family. The details of OFDM and smart antennas are left for the next chapter on WiMAX where you can find much more information on these.

LTE has a goal to reach downlink speeds of 100 Mbit/s and 50 Mbit/s uplink with a latency (one way terminal to Node B) of 10 ms. This sounds impressive – but this is more a burst rate than a sustained (IP level) transfer rate. If you remember that HSDPA can achieve 7.2 Mbit/s burst rate at the physical layer then this looks more realistic. In order to achieve this LTE will utilise several physical layer techniques:

- Scalable bandwidths of 1.25, 2.5,5, 10 and 20 MHz;

- Smart antennas for the downlink (4×2, 2×2, 1×2 and 1×1[17]) and uplink (1×2 and 1×1);

- OFDM (Orthogonal Frequency Division Multiplexing) in the downlink;

- SC-FDMA (Single Carrier Frequency Division Multiple Access) in the uplink;

- Very short transmission time slot (TTI) (0.5 ms);

- Fast PHY Layer HARQ in both down and uplinks;

- Adaptive modulation in downlink (QPSK, 16QAM, 64QAM);

- Adaptive modulation in uplink (BPSK, QPSK, 16QAM).

---

[17] NxM denotes N elements at the base-station (transmitter) and M at the terminal receiver.

The 100 Mbit/s downlink speed refers to a 20 MHz spectrum allocation with OFDM and smart antennas – all of which (including wider spectral bands) tend to increase the spectral efficiency. One of the key goals of LTE was to move away from the fixed 5 MHz carriers of UMTS. In fact LTE is planned to be three to four times more efficient in the downlink than Release 6 HSDPA and two to three times more efficient in the uplink. Since these figures refer to HDPA with type 1 receivers the spectral efficiency gain of LTE over an HSDPA downlink with more advanced receivers will only be a factor of 2 or so. This gain will come from a switch to OFDM in the downlink with smart antennas. OFDM is described in great detail in the next chapter on WiMAX and so there is no point in going over why it is more efficient than CDMA – but most of the factor of 2 efficiency gain comes from the use of smart antennas – which CDMA systems don't benefit from to any great extent due to their wideband nature.

In the uplink it is worth saying a few words about SC-FDMA – as, unlike WiMAX, LTE is not using OFDM in the uplink. The major disadvantage of OFDM, certainly for mobile terminals, is that it requires a high Peak to Average Power Ratio (PAPR) –this happens because the individual signal modulations on the many orthogonal carriers occasionally peak together requiring a very high transmit power. The problem comes in the use of a power amplifier just before transmission – as any hifi buff will tell you that amplifiers are only linear (i.e. no distortion and sound good) when they are operating at a fraction of their maximum power. Unfortunately for OFDM the amplifier needs to be operating in a linear region even at peak transmission levels – so the average operating point has to be backed-off a long way down in the linear region – where the power efficiency of the amplifier is poor: 20% or so. In GSM, by contrast, the PAPR is much lower and the average operating point of the power amp can be set much higher – yielding efficiency nearer 70%. For handheld terminals this can make a huge difference to the power consumption of the terminal.

The LTE uplink gets around this by using SC-FDMA – which is just like OFDM with orthogonal frequencies but, instead of each terminal using all the sub carriers, only a few (different) carriers is used by each terminal. This greatly reduces the PAPR and, hence, the battery consumption. The trade-off, however, is that multipath is no longer eliminated (since the 50Mbit/s uplink is now confined to a narrower bandwidth) and adaptive equalisation is needed at the base-station. This is not such a disadvantage as a complex, power-hungry equaliser at the base-station is cheap and has no implications on the handset battery life. Another interesting feature of LTE is that it is expected to be an FDD system with paired spectrum (although support for TDD is also included). This is the reason why beam forming has been excluded from the standards, since this works best in TDD systems where the uplink (say) can be used to accurately estimate the downlink channel. Only diversity (n × 1 or 1 × n) or downlink MIMO (Multiple In Multiple Out antennas 4 × 2 and 2 × 2) are being developed.

So LTE might have lower battery consumption and slightly higher data rates than WiMAX. What other differences are there? Firstly, LTE is targeting a very low latency – 10 ms round trip time between the terminal and the (evolved) eNode B. This is achieved by using a very short TTI (0.5 ms) as well as a flat architecture (see below) whereby the RNC and Node B are effectively combined in an eNode B. Set up times from idle are also expected to be very low at 65–95 ms. LTE will also ditch the complicated soft handover of CDMA and offer hard handover only. This is negotiated directly between the old and new eNode Bs, eliminating the RNC needed for HSPA hard handover. In addition packets are tunnelled from the old to the new base-stations – giving a service interruption target of 30 ms or so – adequate for VoIP.

LTE standardisation in 3GPP is effectively split into two parts – with the RAN working group looking at the air interface and radio and the Systems Architecture group is tackling the all-IP network (known as the SAE – Systems Architecture Evolution – standards attendees don't have much imagination!). Work of SAE is not as far advanced as in the RAN work and many more details still remain to be decided at the time of writing (Q2 2008). The LTE core is all-IP so there is no circuit switched (CS) part and all services (including voice) are delivered over the packet core. The intention is also that SAE becomes the core for non 3GPP access technologies – such as WiMAX and WLAN with interfaces for these technologies to connect. Compared to a UMTS core the LTE SAE (Figure 4.25) is compressed with the Node B and most of the RNC functionality merged to create an evolved eNode B – this is aided by the removal of soft handover. All of the core network functions go into an "Access Gateway" that actually consists of several logical elements (Figure 4.25):

• Mobility Management Entity (MME) that takes the control plane functions of the SGSN;

• User Plane Entity (UPE) and 3GPP Anchor that take some RNC functions (e.g. the PDCP), the SGSN user plane functions and the GGSN functionality. Note the split between the UPE and 3GPP anchor is not yet decided.

• SAE Anchor – that provides access for non 3GPP systems – such as WLAN.

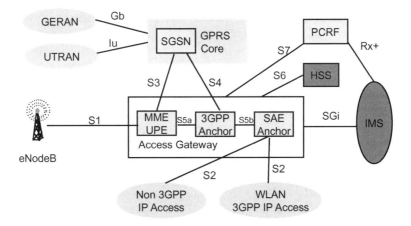

Figure 4-25.   LTE network architecture. (Source: Author. Reproduced by permission of © Dave Wisely, British Telecommunications plc.)

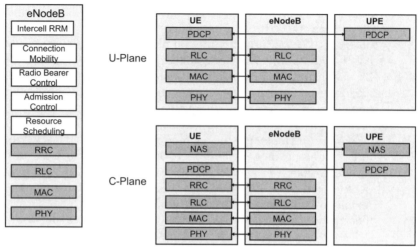

PDCP – Packet Data Convergence Protocol
NAS – Non Access Stratum (eg DHCP)

Figure 4-26.   LTE stack protocols. (*Source:* Author. Reproduced by permission of ©
Dave Wisely, British Telecommunications plc.)

Figure 4.26 shows how the (hopefully by now) familiar protocols are
distributed between the terminal, eNB and UPE/MME. The NAS (Non-Access
Stratum) is basically the IP layer from the gateway (IP address, QoS etc.).

The whole point of SAE is to lower network costs and complexity. This has
been achieved, firstly, by removing the need for soft handover – which itself
allows the network to be "flattened" with the effective removal of the RNC.
Secondly, all transport is now IP – so the *circuit switched* (CS) core has
disappeared. In terms of backhaul the expensive E1 circuits required to support
ATM are replaced by Ethernet transporting IP.

Other features of LTE include support for operating as a single frequency
network to support MBMS (Multimedia Broadcast Multicast Service) – this is
explained in Chapter 7 but is the cellular industry's solution for broadcast
services such as TV. Basically all base-stations have to be fully synchronised
and devote a portion of their resources for broadcast services. In the case of LTE
this will allow the use of IP multicast. If you are particularly interested in the MAC
and PHY details of LTE then Ref. 22 gives an overview and the standards are listed
at the end of the chapter (don't forget you also need to get your anorak cleaned!).
In general they are very similar to those in HSPA and 3G. Resources are allocated
by a dynamic scheduler in the eNode B – these can be long-lived or dynamic
and, for the uplink, resemble the resources maps of WiMAX save for the fact that
only a fixed number of sub-channels are allocated in SC-FDMA in the uplink.
Security reuses much of the 3G SIM based system and QoS is provided by
terminal and eNode B schedulers as in HSDPA. All services are provided by the
IMS with gateways to non IP networks (see Chapter 6 on the IMS).

	HSPA (Rel.6)	LTE
Downlink Peak Rate Mbit/s	14.4	144
Downlink Spectral Efficiency (Bit/s/Hz)	0.75	1.84
Uplink Peak Rate Mbit/s	5.7	57
Uplink Spectral Efficiency (Bit/s/Hz)	0.26	0.67

Table 4-6. Results of modelling the various phases of HSPA and LTE (LTE Forum).

So, having had a (very rapid) tech overview of LTE the first question is probably "will it work"? There is certainly some evidence that burst data rates will approach 100 Mbit/s (at the PHY layer) in the downlink with a 20 MHz channel allocation. Table 4.6 (LTE Forum) shows the results of modelling HSPA and LTE – again they offer, I think, indicative and comparative, rather than definitive answers but you can see how LTE stacks up with HSPA. Ref. 23 gives some further published rates for LTE modelling.

Is LTE Fourth Generation Mobile? Well it certainly fits well with the evolution of the previous generation in that it is a totally new air interface, will appear a decade after the previous generation and will offer something like five times improvement in general performance (e.g. data rate). On this measure it probably should be termed 4G.

However, some within the industry see 4G as synonymous with "IMT Advanced". The timeline for IMT advanced deployment is roughly 2015 with a headline data rate of 100 Mbit/s targeted. Indeed LTE has now been registered with the ITU as a 3G technology and something called "LTE advanced" has appeared as a study item to represent 3GPP's IMT advanced (i.e. 4G) technology offering.

Some of this posturing is about managing spectrum costs: as Adrian Scrase said at a recent LTE conference (Ref. 24) "Don't fuel the licensing debate by improper use of the term 'Generation'" and industry insiders continue to use the term 3.9G. So, maybe, we can settle on 3.9G for LTE? For an update on possible spectrum for LTE have a look at the spectrum section of Chapter 2.

It is legitimate to ask what users will do with 100 Mbit/s on the move? The author has a laptop with a crack in the back of its screen – sustained whilst attempting to see how far down the street a WLAN would transmit a streamed DVD! Unless screen technology improves – with something novel such as retinal scanning or immersive 3D visualisation – how will 100 Mbit/s be used? It is now possible to buy 1 TByte hard drives for £300 or so, Flash memory cards up to 50 Gbyte are also becoming available and will start to appear in mobile devices. Given that a typical film takes about 100 Mbyte storage for replay on a mobile device – it would be possible to cache the top 500 films on a 50 Gbyte flash card (and in five years 10 000 films with miniature hard drives). With such caching facilities and also the rise of short range, high bandwidth, access technologies – such as WLAN and Ultra Wide Band – which can "top up" and refresh such prodigious caches then it is hard to envisage a real use for these very high data rates. Maybe homes will go completely wireless and all services – including HDTV will be delivered over cellular? Sadly there is not enough capacity (Ref. 25) even for LTE to convey IPTV to a large number of homes.

Finally, a quick summary of the position for EVDO afficionados – what is the LTE equivalent for a CDMA2000 network? The answer is EVDO Rev.C which uses... Yes you guessed it... OFDMA with MIMO and an all-IP network. Key aims of Rev.C are:

- Scalable solution capable of deployment in existing cellular and PCS spectrum;
- Major improvements in VoIP efficiency (over Rev.B);
- Support for higher bandwidth symmetric services – including video telephony;
- MIMO antennas to increase up and down link data rates;
- Use of OFDM;
- Sectorisation techniques using smart antennas (e.g. Spatial Division Multiple Access – see WiMAX Chapter 5).

Work started in 3GPP in March 2005 and two camps quickly emerged: Strictly Backward Compatible to allow Rev.A/B mobiles to share carriers with Rev.C and Loosely Backward Compatible – requiring Rev.A/B and Rev.C users to be on different carriers whilst allowing a reuse of much of the mobile stack. In the end the loose camp prevailed. Key differences of Rev.C to LTE are the use of noncontiguous spectrum (to allow odd chunks to be used) and a hybrid CDMA-OFDM uplink. CDMA is used for control messages (fast access, power control and handover) and (optionally) for low latency, low rate traffic such as VoIP and gaming. Handover is also complicated as up and down links may be to different base-stations! Ref. 26 gives the following performance improvement with Rev. C (Table 4.7).

At a recent conference (Ref. 28) it was suggested that actually the evolution path for EVDO would be more likely to be LTE – which would create further unification in the diverse 3G standards.

	Rev. 0 (1.25 MHz)	Rev. A (1.25 MHz)	Rev. B (5 MHz)	Rev. C (20 MHz)	Rev.C (20 MHz) + MIMO
Downlink Mbit/s	2.4	3.1	14.7	70	260
Uplink Mbit/s	0.153	1.8	5.4	70	

Table 4-7. EVDO RevC possible performance (Ref. 26).

4.9 Conclusion

3G was originally marketed as bringing in lots of whizzy new services: video calling, football highlights, messaging and so on. Then the Internet happened and it was re-branded the Mobile Internet. Sadly its design pre-dated IP and it was soon obvious that 3G was much better suited to cheap voice. Technical limitations of the radio interface meant long set up times, high latencies and low data rates and these have held back Internet-style services – such as VoIP, IM and email. 3GPP and 3GPP(2) have recognised this and delivered to the industry a series of upgrades that attempt to improve all these aspects of the mobile data experience. Crucially for operators they are cheap – a few $ per customer (especially with handsets churning every 18 months or so) – and evolutionary (with each upgrade building on the last and fully backwards compatible).

The performance of HSPA and EVDO Rev A/B is now starting to reach the level provided by fixed broadband – in terms of cost as well as data rate. For broadband users with small appetites then the option of using mobile networks for data as well as voice is becoming attractive. So attractive that 18% of Austrians did just that in 2007 and the incumbent fixed operator reduced prices by some 40%. Of course there is a limit – the current networks are not going to support file sharing and IPTV – there just isn't enough capacity. A whole lot of new spectrum and more efficient technologies will be needed for that transition.

The key for the Mobile Operators is whether they are able to continue selling all this extra capacity as services that are brought directly from the operator – like music downloads, video calling, rich VoIP services – or whether they simply sell it per Mbyte as broadband currently is and the likes of Google, Visa and Amazon hoover up all the value add. The answer to this question is also at the heart of what operators will do with technologies like WiMAX and LTE that move them further towards broadband and further from circuit voice. In the wider scheme of things these technologies are not 4G – I think the term 3.5G is about right – since they give us what 3G promised all those years ago.

References

[1] BT measurements in summer 2007.
[2] http://www.3g.co.uk/PR/April2005/1384.htm – reports IPR cost approaching 25–30% of handset selling price costs.
[3] Tomi Ahonen – well known character on the mobile conference circuit – excellent website with much information http://www.tomiahonen.com/.
[4] "Network performance of mixed traffic on high speed downlink packet access" and dedicated channels in WCDMA and, K.I. Pedersen *et al.* IEEE Proc. Vehicular Technology Conference, Sept. 2004.
[5] "UMTS Networks", H. Kaaranenn *et al.*, John Wiley & Sons Ltd, 2001, ISBN 0471 48654 X.

[6]   "HSDPA/HSUPA for UMTS: High Speed Radio Access for Mobile Communica-
      tions (Hardcover)" by Harri Holma and Antti Toskala (eds), John Wiley & Sons Ltd,
      2006. ISBN-13: 9780470018842.
[7]   Heath, M., Brydon, A., Pow, R., Colucci, M. and Davies, G.: Prospects for the
      Evolution of 3G and 4G, *Analysis* (2006).
[8]   "The Evolution of UMTS/HSPA Around the World", Alan Hadden,President,
      GSA, Global Mobile Suppliers Association, Informa Telecoms & Media LTE
      World Summit Berlin, 21–23 May 2008.
[9]   "Wireless Media Devices", Strategy Analytics, 2008.
[10] "Mobile Broadband: EDGE, HSPA, LTE", Peter Rysavy, 3G Americas white paper,
      2006,   available   at   http://www.itu.int/ITU-D/imt-2000/TechnicalArticles/
      2006_Rysavy_Data_Paper_FINAL_09.15.06.pdf.
[11] "The Evolution of UMTS/HSPA Around the World", Alan Hadden,President,
      GSA, Global Mobile Suppliers Association, Informa Telecoms & Media LTE
      World Summit, Berlin, 21-23 May 2008.
[12] 3G Americas press release 13 June 2006.
[13] "The Evolution of UMTS/HSPA Around the World", Alan Hadden,President,
      GSA, Global Mobile Suppliers Association, Informa Telecoms & Media LTE
      World Summit, Berlin, 21–23 May 2008.
[14] "Using SIP-based Applications and, SIP Transport for Media Independent Hand-
      over", Vijay Kesavan, Christian Maciocco, Andy L.Y. Low and Dave Wisely SIP
      Forum, Paris, Feb. 2008.
[15] "Mobile Broadband: EDGE, HSPA, LTE", Peter Rysavy, 3G Americas white paper,
      2006 – EDGE evolution p. 20.
[16] See   http://www.cdg.org/worldwide/index.asp?h_area=0&h_technology=999
      for current CDMA2000 deployments.
[17] "Measured TCP Performance in CDMA 1x EV-DO Network", Youngseok Lee,
      Proceedings of the Passive and Active Measurement Conference pamconf.org
      available at: http://www.pamconf.org/2006/papers/s1-lee.pdf.
[18] "VoIP over cdma2000 1xEV-DO revision A", Yavuz, M., Diaz, S., Kapoor, R.,
      Grob, M. Black, P., Tokgoz, Y. and Lott, C. *Communications Magazine*, IEEE, Feb.
      2006, Vol 44, No. 2, pp. 50–57.
[19] www.**Femtoforum**.org/.
[20] Sprint — http://www2.sprint.com/mr/news_dtl.do?id=12960.
[21] Next Generation Mobile Network — http://www.ngmn-cooperation.com/.
[22] "Technical overview of 3GPP LTE", H.G. Myung,Feb. 2007 – available at http://
      hgmyung.googlepages.com/3gppLTE.pdf.
[23] "The 3G Long-Term Evolution – Radio Interface Concepts and Performance
      Evaluation", Erik Dahlman, Hannes Ekström, Anders Furuskär, Ylva Jading, Jonas
      Karlsson, Magnus Lundevall, Stefan Parkvall,VTC Spring 2006 – available at
      http://www.ericsson.com/technology/research_papers/wireless_access/doc/
      the_3g_long_term_evolution_radio_interface.pdf.
[24] US spectrum auction – see http://business.timesonline.co.uk/tol/business/indus-
      try_sectors/telecoms/article2882635.ece.
[25] See Heath, M., Brydon, A., Pow, R., Colucci, M. and Davies, G. "Prospects for the
      Evolution of 3G and 4G", (2006).
[26] S. Vasudevan Air Interface Evolution in 3GPP, 3G LTE Conference, The Café
      Royal, London, UK (October 2006).
[27] UMTS Networks, H. Kaaranen et al., , 2001, ISBN 0471 48654 X.
[28] Global Mobile Suppliers Association, Informa Telecoms & Media LTE World
      Summit, Berlin, 21–23 May 2008.

More to Explore

General
"Future Mobile Broadband: HSPA, EV-DO, WiMAX & LTE" – Mike Roberts – Informa Telecom 2006.

HSDPA
Book "HSDPA/HSUPA for UMTS: High Speed Radio Access for Mobile Communications (Hardcover)" by Harri Holma (Editor), Antti Toskala (Editor), John Wiley & Sons Ltd, 2006. ISBN-13: 9780470018842. Difficult in places as authors are not native English speakers and the fluency suffers, but all the details on HSPA. Best HSPA book.

Tutorial http://www.3g4g.co.uk/Tutorial/ZG/zg_hsdpa.html

3G Americas report on deployment of HSPA
http://www.3g4g.co.uk/Lte/Tutorials/3GAmericas_UMTS_Rel7_WP.pdf

Nokia Whitepaper
http://www.nokia.com/NOKIA_COM_1/About_Nokia/Research/Demos/
    HSDPA/HSDPA_A4.pdf

Book chapter on HSDPA – out of date (2002) but available free online: http://
    www.sis.pitt.edu/~dtipper/hsdpa.pdf

CDMA2000 Evolution
Excellent site on deployments and news for CDMA2000
http://www.cdg.org/technology/3g_1xEV-DO.asp

Nokia Whitepaper on CDMA2000 evolution
http://nds2.ir.nokia.com/NOKIA_COM_1/About_Nokia/Press/White_Papers/
    pdf_files/evdvwp.pdf

LTE
Standards freely available at: http://www.3gpp.org/ftp/Specs/archive/
    25_series/

Requirements for Evolved UTRA (E-UTRA) and Evolved UTRAN (E-UTRAN)-
    http://www.3gpp.org/ftp/Specs/html-info/25913.htm
Evolved Universal Terrestrial Radio Access (E-UTRA); Long Term Evolution
    (LTE) physical layer; general description. http://www.3gpp.org/ftp/Specs/
    html-info/36201.htm

Physical layer aspect for evolved Universal Terrestrial Radio Access (UTRA)
    http://www.3gpp.org/ftp/Specs/html-info/25814.htm

See also: http://www.3gpp.org/Highlights/LTE/LTE.htm

160   IP for 4G

Online resources:

"Technical overview of 3GPP LTE", H. G. Myung, Feb. 2007 – available at http://hgmyung.googlepages.com/3gppLTE.pdf

"3GPP LTE & 3GPP2 LTE Standardization" by Dr Lee, Hyeon Woo at Samsung Electronics (June 2006) – available at http://www.krnet.or.kr/board/include/download.asp?no=30&db=program&fileno=2

"3GPP Long-Term Evolution/System Architecture Evolution: Overview" by Ulrich Barth at Alcatel (Sep. 2006) available at: http://www.ikr.uni-stuttgart.de/Content/itg/fg524/Meetings/2006-09-29-Ulm/01-3GPP_LTE-SAE_Overview_Sep06.pdf

Papers on LTE

H. Ekström, A. Furuskär, J. Karlsson, M. Meyer, S. Parkvall, J. Torsner and M. Wahlqvist, *IEEE Commun. Mag.*, vol. 44, no. 3, March 2006, pp. 38–45.

E. Dahlman, H. Ekström, A. Furuskär, J. Karlsson, M. Meyer, S. Parkvall, J. Torsner and M. Wahlqvist, *Ericsson Review*, no. 2, 2005.

E. Dahlman, H. Ekström, A. Furuskär, Y. Jading, J. Karlsson, M. Lundevall and S. Parkvall, *IEEE Vehicular Technology Conference (VTC) 2006 Spring*, Melbourne, Australia, May 2006.

UMTS Forum www.umts-forum.org Report 41 studying LTE.

Chapter 5: WiMAX

5.1 Introduction

It is hard to read anything about cellular mobile without coming across predictions and headlines that WiMAX is going to revolutionise mobile data – either by dramatically increasing data rates ("It has a service range of up to 50 Km and provides data rates of up to 280 Mb/s per base station") or reducing costs to residential broadband-like levels (Ref. 1). Some have hailed it a 4G technology – including Sprint Nextel in the US choosing it as its next generation mobile data technology and promising a roll-out in the next two to three years.[1] There are other reasons to get excited about WiMAX. Firstly, it uses a new-fangled physical layer technique called OFDMA (Orthogonal Frequency Division Multiplexing Access) – splitting the spectrum available up into hundreds of orthogonal (independent) carriers and then adjusting each carrier to the exact radio propagation on that carrier. Secondly, WiMAX systems can easily employ smart antennas – which really means multiple antennas – at the receiver, base-station or both. Smart antennas are able to get away from 360 degree omni directional transmissions and target much narrower beams. In certain circumstances smart antennas – with M transmitters and M receivers – can achieve up to an M-fold increase in transmission rates.

---

[1] "Sprint Nextel has created a unique business model designed to foster the rapid deployment and adoption of mobile WiMAX technology in the United States and abroad. Sprint Nextel is expecting to invest $1 billion in 2007 and between $1.5 billion and $2 billion in 2008 relating to the 4G mobile broadband network. The WiMAX technology to be deployed in the network is expected to offer a cost-per-megabit and performance advantage that reflects a substantial improvement in the comparable costs for the current 3G mobile broadband offerings" – *Sprint News* Ref. 2.

IP for 4G   David Wisely
© 2009 John Wiley & Sons, Ltd.

WiMAX is also exciting because it comes not from the usual mobile standards bodies (3GPP and 3GPP2) but from the IEEE. The IEEE is much more a computer industry body – with close links to computer chip vendors and industry players. The leading proponents of WiMAX are companies such as Intel and Samsung – whilst traditional mobile industry players have been more lukewarm towards WiMAX. One of the exciting things about WiMAX is that it might integrate better with IP technologies and be better suited than 3G for IP data applications. As we have seen, 3G – with its need for an ATM transport technology and long latency and even longer data transfer session set up times – is not well matched for IP applications. The inference is that the low costs, open standards, limited IPR cost and the scale of the Internet will be carried over into WiMAX as an antidote for the proprietary, limited interoperability and high IPR licensing costs of 3G. Table 5.1 compares WiMAX with the 3G evolutionary approach.

Finally there is the view that WiMAX is going to be "WLAN on steroids" (Ref. 3). All the good things about WLAN – high bandwidth, low-cost, small coffee shops, real email, IM and web browsing without restrictions, time and

	3G evolution	WiMAX
Standards body	3GPP and 3GPP2	IEEE 802.16
Background of supporters	Cellular mobile	Computer industry
Major developers	Mobile equipment manufactures, mobile operators	Computer chip manufacturers
Timescale	2006–2008 for 3.5G – 2011–2013 for next generation (LTE)	2007–2009 for fixed WiMAX (802.16) and 2008–2011 for mobile WiMAX (802.16e)
Early adopters	Existing mobile operators	New entrants
IP support	Only in next generation (LTE) – 3.5G has better IP performance but is not "all-IP" and voice is carried over circuits.	All-IP system with all voice carried as VoIP. IP QoS and mobility management.
Terminals	Phones, PDAs and PCMCIA cards	Built in to laptops and PDAs. Phones expected to emerge.
Spectrum	Existing cellular spectrum but calls for significant new spectrum for LTE.	3.5GHz for fixed WiMAX and 2.3–2.6GHz for Mobile WiMAX. Calls for lower frequencies to be allocated.
Capabilities	Full cellular system with support for: Mobility, QoS, security, handover, voice and charging.	Full cellular system with support for: Mobility, QoS, security, handover, voice and charging.

Table 5-1. comparison of the background between WiMAX and 3G evolution.

not volume-based charging, user controlled choice of operators, SIM-free operation . . . will be adopted. In addition all the advantages of cellular mobile will be added – mobility, wide area coverage, handover to other technologies, strong security and guaranteed QoS.

To what extent WiMAX is a 4G system is a moot point and I will return to this at the end of the chapter. In the meantime this chapter will start with a description of WiMAX in all its gory detail – PHY and MAC layers and network architecture – including a detailed (but non mathematical) description of the key technical innovations. Firstly, though, we need to look at the history of WiMAX and separate the two flavours of WiMAX – fixed and mobile.

5.2 Overview of WiMAX History

802.16 standards are quite confusing – when the group started in 1999 it was divided into two groups 802.16a (approved Jan. 2003) (for frequencies 2–11GHz) and 802.16 (approved December 2001) (10–66GHz). Only the 802.16a standard is used for WiMAX. Both (as well as various historical releases) are consolidated in the 802.16-2004 standard. This is also referred as 802.16d – after the 802.16d project that contributed to part of the standard.

In December 2005 new amendments to the 802.16-2004 standard were agreed – adding support for mobility, smart antennas and a new multiplexing scheme for multiple users – OFDMA (Orthogonal Frequency Division Multiple Access). This is known as 802.16e.

In the same way that the WiFi alliance (Chapter 3) has been responsible for testing, certifying and interoperability of WLAN products against the 802.11 standard, so the WiMAX forum was created in 2001 to play a similar role. It was here that the term WiMAX (Worldwide Interoperability for Microwave Access) was coined. WiMAX certification is a combination of testing that the products conform to the standard and are able to interoperate. Certification of 802.16-2004 (fixed) equipment began in 2006 and the first certification of 802.16e equipment is expected in 2008 (for a list see Ref. 4).

In Korea a similar OFDM-based broadband mobile standard emerged and was published in 2004 with three licences assigned in January 2005 and commercial services launched in June 2006 – Figure 5.1 shows a WiBro handset, as the standard came to be known. Fortunately for the sake of unity it was possible to make changes to the 802.16e standard to include WiBro and make it possible for it to be certified as WiMAX.

The 802.16 standard defines the concept of a profile – essentially a combination of:

- MAC scheme;
- PHY layer type;
- TDD or FDD;
- Frequency band;
- Power class.

Figure 5-1. WiBro terminal. (Source: BT. Reproduced by permission of © British Telecommunications plc.)

Fixed			
Frequency Band (GHz)	Duplexing mode	Channel bandwidth (MHz)	Profile Name
3.5	TDD	7	3.5T1
3.5	TDD	3.5	3.5T1
3.5	FDD	3.5	3.5F1
3.5	FDD	7	3.5F1
5.8	TDD	10	3.8T
Mobile			
Frequency Band (GHz)	Duplexing mode	Channel bandwidth (MHz) and (number of carriers)	
2.3–2.4	TDD	5 (512) 8.75 (1024) 10 (1024)	
2.305–2.320 2.345–2.360	TDD	3.5(512) 5 (512) 10 (1024)	
2.496–2.690	TDD	5 (512) 10 (1024)	
3.3–3.4	TDD	5 (512) 7 (1024) 10 (1024)	
3.4–3.8 3.4–3.6 3.6–3.8	TDD	5 (512) 7 (1024) 10 (1024)	

Table 5-2. Fixed and Mobile WiMAX certification profiles (note that does not imply these frequencies are allocated for WiMAX).

The WiMAX forum has defined a number of profiles for both fixed and mobile WiMAX (Table 5.2).

The first part of the chapter will first look at OFDM – the first "magic bullet" of WiMAX – followed by a description of the MAC and PHY layers of Mobile WiMAX 802.16e. To save time this section will concentrate on the mobile WiMAX standard and, in fact many commentators expect all chips to be produced to this standard – even for fixed applications due to the potential economies of scale. Then we will tackle the key mobility additions to WiMAX – handover, QoS and overall network architecture.

5.3 OFDM

Fixed WiMAX uses OFDM as part of its physical layer and mobile WiMAX uses a variant – OFDMA – which combines OFDM with a MAC function. OFDM is in fact all the rage in wireless systems – found in the latest WLANs (802.11a and 802.11g – see Chapter 3) and broadcast technologies (such as DVB-T – see Chapter 7). OFDM is important because it potentially offers a solution to the (dreaded) multipath problem that is particularly troublesome for cellular systems in the frequency range 0.5–3GHz (where radio signals take a number of different paths between the transmitter and receiver). Multipath limits the performance of W-CDMA systems such as 3G: OFDM, in theory at least, offers greater potential throughput for a given amount of spectrum.[2] The other important thing about OFDM is that it is very well matched to the use of smart antennas in a way that CDMA systems are not – providing further (potential) advantage. You might be forgiven for thinking OFDM was a recent great technical breakthrough that had burst upon the scene: in fact it has been kicking about for donkey's years (my Dad had heard of it and he is so old that English Heritage have made me an offer for him[3]!!) – it was even considered as a technology for GSM as well as 3G. What has really changed is that Moore's law has finally allowed the complicated multi processor electronics needed to execute the processing-hungry algorithms to be made at a cost, size and power consumption comparable with GSM.

In a "traditional" radio system – let's say GSM – the signal from the transmitter is sent to the receiver by modulating a radio carrier – the carrier in this case is 200KHz wide (but each voice conversation gets only 1 of 8 TDMA time slots). A modulation is chosen – say BPSK (Binary Phase Shift Keying – phase "up" and phase "down") and the entire carrier 200KHz is modulated (its actual frequency

---

[2] Called spectral efficiency and expressed in bit/s/Hz/cell – in R99 it reaches about 0.2 bits/Hz/cell and 1 bit/Hz/cell in HSDPA.

[3] Sorry Dad. In fact it was first proposed by Bell Labs in 1966 and adapted for mobile in 1985.

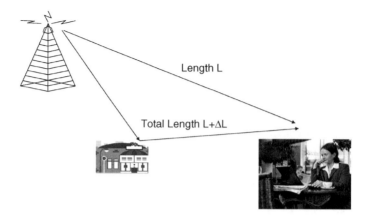

Figure 5-2. Origin of fading. (Source: Author. Reproduced by permission of © Dave Wisely, British Telecommunications plc.)

range is something like 900.4MHz to 900.6MHz). The signal from the transmitter typically "bounces" around the environment between it and the receiver (imagine you are in a ground floor office and the transmitter is ½ a mile away with no line of sight – that is a fairly typical cellular environment). The actual propagation path is not unlike optical rays – some bounce off the building opposite, some off a passing bus and so on. At the receiver you end up with a number of time-delayed versions of the signal from the transmitter – each with a different delay (i.e. phase) and strength – typically the delay spread is 0.1 to 3 microsec.[4]. This is the multipath problem of cellular radio and is often characterised by a delay spread figure (the delay spread is the root mean square [a fancy average!] of all the delayed components weighted by power). Multipath causes a couple of fundamental radio problems that are at the heart of why OFDM is "magic".

The first "problem" of multipath is fading – the constructive and destructive interference of the different radio "rays" causing the signal level to rapidly fluctuate. Fading is a very serious problem for many radio systems (but not for most mobile systems as will be explained in a minute) – with deep fades being of up to 30 dB being common. It is important to realise that the fading is taking place before the signal is in anyway demodulated and happens even without any data modulation. It is easy to see this if you image a single tone – i.e. a pure sine wave – at 900MHz (say). If we have just two rays (Figure 5.2) then the interference will be a constructive maximum when $\Delta L = n\lambda$ where ΔL is the path difference, n and integer and λ the wavelength. It will be a destructive minimum when $\Delta L = (n + 1/2)\lambda$. If the two rays have equal power then the interference will be total. If, say, one has a quarter the power (half the amplitude) of the other, then the difference between the maximum and

---

[4] Indoor micro-cell 0.001–0.05 μs. Open area <0.2 μs. Suburban macro-cell <1 μs. Urban macro-cell 1–3 μs. Speed of propagation 3×10^8 metres per second therefore 1 μs delay spread variation corresponds to 100 metres variation in path length (1 ns per foot).

minimum received power will be a factor of 9 – almost 10 dB variation. Moreover this variation will happen when the terminal moves a distance of ½ a wavelength. At 1 GHz the wavelength is 30cm. In practice many rays with different powers converge on the terminals and the signal variation can be anywhere from 10 dB to 30 dB depending on the frequency and radio environment. Typical fading numbers are 10 to 20 dB for GSM in urban environments.

Now imagine I repeated the above experiment but I was able to vary the frequency being used – what would happen? Well what you would see was that as you gradually changed the frequency the received power would cycle between a maximum and a minimum. It is fairly simple to see this – if λ_1 is the starting wavelength then we have (for constructive interference) $\Delta L/\lambda_1 = n$. If we increase λ_1 to λ_2 to cycle through destructive interference to the next constructive maximum we must have $\Delta L/\lambda_2 = n + 1$. A simple piece of manipulation (remembering $c = v\lambda$ – where c is the speed of light and v the frequency) gives $\Delta v = c/\Delta L$ or $\Delta v = 1/DS$ where DS is the delay spread of extra time it takes to travel the longer path. This is in fact true when there are a whole bunch of rays with different strengths. So if the delay spread is 1 μsec then the frequency needs to be changed by 1 MHz to average over destructive and constructive interference.

The answer to fading, then, is to send the signal over a range of frequencies – bigger than the inverse of the delay spread – and then there will be an average of constructive and destructive interference. Some spreading takes place anyway when modulation is applied to a signal. If I am transmitting with a pure 900 MHz carrier and then modulate it with 1 Mbit/s of data (using just amplitude modulation, say) then the modulated signal will spread over a range of frequencies but most of the power will be between 899 MHz and 901 MHz. This would be enough to reduce the fading margin and this would be a wideband system (meaning the spread of frequencies due to modulation is larger than the delay spread). W-CDMA, which spreads even a 12 kbit/s voice call into a whole 5 MHz, is an extreme example of a wideband system.

Wideband spreading is an excellent solution to fading but there is a catch! – the received signal is distorted. You can see why it is distorted by thinking about either the frequency or time domains. If you take any signal (e.g. a piece of music) and alter the amplitudes of the frequency bands (e.g. fiddling with the treble and bass controls on your hifi!) then you get distortion. In the same way a wideband signal transmitted over a multipath channel is received with a distortion since the response of the channel (the aggregate power and phase received from all the rays) varies with frequency (the very property we desire to mitigate fading) – moreover it varies across the frequency range of the desired signal.

In the time domain it is also easy to see how this distortion – called InterSymbol Interference (ISI) arises (Figure 5.3). Multiple time-shifted copies of the same digital sequence now arrive out of time alignment at the receiver. Remember we are now looking at the underlying modulation of the carrier – a wideband system has a bit period (strictly a symbol period) shorter than the delay spread – so that after demodulation the bits are misaligned in time from

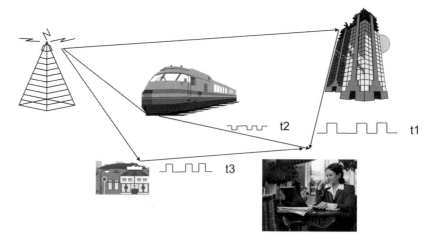

Figure 5-3. *Inter Symbol interference (ISI)*. (*Source:* Author. Reproduced by permission of © Dave Wisely, British Telecommunications plc.)

the different rays. If you could resolve the rays in space you could use all of them without distortion but in a standard receiver they are all mixed – with their different amplitudes and phases to create ISI. Say the sequence is 00100101 – each constituent "ray" or radio path is delayed sufficiently for some of the copies of the first one to arrive in the following 0 slot and some (usually reduced) to arrive in the next slot (Figure 5.3). Now the symbol period is short compared to the delay spread – that is a wideband system. This is just the time domain result of applying frequency domain filtering – because some of the frequencies are in fade and some constructively interfere and are received with a high signal level.

The traditional solution to ISI is to use an equaliser – as GSM does. An equaliser uses a special signal ("training sequence") with known frequency content from the transmitter to estimate the frequency response of the channel and compensates for it. This is analogous to recording a record with the wrong bass and treble settings – if you know what part of the song should sound like you can estimate (e.g. by trial and error) the right treble and bass settings needed to make the song sound OK. Equalisers are, however, complex and imperfect and the channel is never stationary – changing rapidly as the terminals move or even as the environment changes (e.g. buses!). Also when the gain has to be turned up you start to hear the underlying noise. Other solutions include resolving the different multipath signals in time (e.g. a CDMA Rake receiver – see Chapter 2) and space (using MIMO – see the section on smart antennas later in this chapter).

Multipath is troublesome for all mobile communications systems at frequencies currently used (at very high frequencies transmission is mostly line-of-sight and multipath is less of an issue). In TDMA systems (like GSM) it manifests itself as Intersymbol Interference as the equalisers are never perfect. In W-CDMA systems (such as UMTS and CDMA 2000) it ruins the orthogonality of the codes in the downlink – the higher the multipath the more interference is caused by orthogonal codes used by other users in the same cell. In fact this is such an issue for CDMA systems that most interference (up to 70%) is generated within the same cell and equalisers (as are planned for HSPA) improve the overall performance especially on the downlink where the restoration of orthogonality reduces the interference level – in the uplink the codes are not orthogonal to start with and so different techniques, such as diversity and interference cancellation are used.

OFDM, on the other hand, eliminates the problem of multipaths completely. It does this by dividing the available spectrum into, typically, 1000 or more sub-carriers, each of which is an independent, narrowband, channel. OFDM is a form of Frequency Division Multiplexing – whereby each user gets a number of frequencies. In OFDM the frequencies – called sub-carriers or tones – are specially chosen to be orthogonal. This is a slightly complex thing to explain. Imagine there are three frequencies – an Irish frequency F1, a Welsh one F2 and a Scottish one F3. If these three frequencies are pure tones – meaning they are just sinusoidal signals that go on for ever – then the ability of any receiver to discriminate between them is determined by the narrowness of its filters. In principle very close tones could be resolved. However, modulate one of those tones with some information – using BPSK, 16 QAM or your favourite modulation scheme – and the bandwidth of the tone broadens by a factor of a few times the bandwidth of the data. To give an example – Taffy radio uses a pure tone of 400MHz to send a broadband signal of 1MBit/s using BPSK (say). The result – when viewed on a spectrum analyser– would be significant power broadcast from about 397MHz to 403MHz and Kilt radio (F3) would need to transmit on 406MHz or so.

OFDM is different – OFDM tones have the property that when you modulate them – although they spread in bandwidth – you can still resolve them from each other – that is why they are called orthogonal! You can in fact perform any linear operation on them and still recover the original signals. Figure 5.4 shows

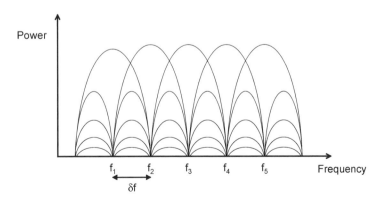

Figure 5-4. OFDM tone power. (Source: Author. Reproduced by permission of © Dave Wisely, British Telecommunications plc.)

the power spectrum of five tones – at the exact centre frequency of F3 – F1, F2, F3 and F4 (broadened due to data modulation) are zero. Imagine using a simple amplitude shift keying (i.e. on or off) to modulate F3 – it is easy to see that the receiver can work out the bit stream of this sub-carrier – even though nearby sub-carriers overlap it (which is why OFDM offers much better spectral efficiency than FDM). The sub-carriers are evenly spaced with spacing δf and, to ensure orthogonally, the symbol period (i.e. how long the modulation is applied for each sub-carrier symbol) is T = n (1/δf) where n is some integer.

OFDM can avoid completely the problem of multipath – the sub-carriers are narrowband because the length of each symbol (be it QAM, BPSK or whatever) is much longer than the delay spread. So although multipath spreads the symbol out the amount is insignificant. If there are N (say 1000) sub-carriers then the symbol rate on each one is only 1/N that which would be needed for a single carrier system. So, to give an example: A system with:

1. A single carrier using 10MHz of spectrum could support a 8 Mbit/s data rate using 16QAM (4 bits/symbol) with a 2 Msymbol/s modulation a 500 ns symbol period).

2. An OFDM system with 1000 sub-carriers each with a rate of 2 ksymbols/s – a 500 microsec symbol period could support the same data rate.

If the multipath delay spread is fairly typical 1 microsec– then it is obvious that the single carrier system is wideband and will require either an equaliser or some other way of combating multipath – such as the CDMA Rake receiver. Each OFDM sub-carrier is clearly narrowband (the symbols are 500 times longer than the delay spread) and don't suffer distortion. They will, however, suffer fading – and the degree of the fade will obviously change if the terminal moves. However, at a given location, some carriers will be in deep fade and some will have a strong signal level. Because there are 1000 carriers it is possible (in principle) to measure the signal-to-noise ratio on each channel and adapt the power and modulation to fit exactly with the current propagation conditions on each carrier (Figure 5.5). Carriers that have favourable propagation characteristics can be given high power allocation and those with poor characteristics (e.g. in a deep fade) can even be turned off. Even better is the fact that every sub-carrier will likely have excellent propagation for some user (if there are enough of them). This is even better than HSDPA – where a single channel of 5MHz is (momentarily) given to a single user. HSDPA is clearly a compromise because 5MHz is (very) wideband and only some of these frequencies can really have excellent propagation between the mobile and the BS (although this is mitigated in HSDPA with advanced receiver types and diversity. In OFDM it is possible to pick the best sub carriers – with the best propagation – and use these. The remaining sub carriers are then used for a second user where they offer better propagation. WiMAX doesn't yet offer this level of sophistication but its sub-channelisation goes some way along these

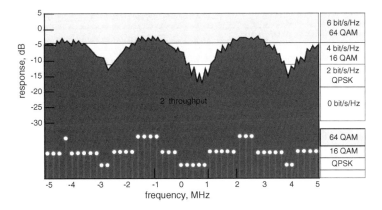

Figure 5-5. Principle of OFDM (courtesy BTTJ). (Source: Author. Reproduced by permission of © Dave Wisely, British Telecommunications plc.)

lines and offers future upgrade possibilities. Mobile WiMAX does offer sub-channelisation which will do this.[5]

The only influence that multipaths has on OFDM is that a cyclic prefix – essentially a "guard" time equal to the multipath delay, must be added to every symbol. This is because multipath causes the different sub-carriers (and reflected/diffracted copies of the same sub-carriers) to arrive with different times at the receiver. For OFDM to work it is necessary to time-align the phase of the different sub-carriers. This can be done easily but only if the symbol is extended in time by at least the maximum delay spread (Figure 5.6).[6] Actually creating an OFDM signal is simple (Figure 5.7) – the data stream is split into N streams,[7] modulated and then an Inverse Fourier Transform operation is performed.[8] The IFT is done digitally on a signal processing chip – it is the development of this – with the right power consumption, performance and price – that has made OFDM practical in the last few years.

The beauty and power of OFDM is that it offers great flexibility in how users are allocated sub-carriers and how sub-carriers are coordinated between neighbouring base-stations. The following are often cited as reasons why

---

[5] There is also a mode – called Partial Use of Sub-Channels (PUSC) – that allows the sub-channels to be divided more finely.

[6] The cyclic prefix is an exact copy of a proportion, say 1/8, (values tend to range from $1/2$ to 1/32 of the symbol time) of the symbol to be transmitted copied from the end of the symbol to the start. This increases the symbol length by 1/8. The receiver starts the Fourier Transform near the end of the cyclic prefix at a point when all delayed signals are expected to have arrived, therefore the delay spread is ignored.

[7] Any number of sub-carriers could be used but using powers of 2, e.g. 1024 sub carriers allow the use of Fast Fourier Transforms – special computing short cuts – as these lend themselves to fast processing on computing platforms.

[8] In a Fourier Transform (FT) a signal is decomposed into orthogonal sine waves and the strength (or modulation) of each sine wave is the output of the transform. For an Inverse FT (IFT) – the components go in and the modulated sine waves come out.

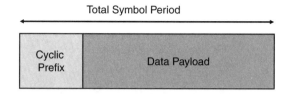

Figure 5-6. Guard band extension to allow orthogonally. (Source: BTTJ. Reproduced by permission of © British Telecommunications plc.)

OFDM has the potential to offer a much higher spectral efficiency for mobile systems than CDMA-based systems:

- The orthogonal nature of OFDM limits intra-cell interference (in CDMA it is reckoned up to 70% of interference is intra cell Ref. 5).

- OFDM systems offer a frequency re-use factor of 1 (with smart antennas).

- The sub-carriers can be used in an optimal way for each user and according to specific propagation conditions (carrier 1 is in fade for user 1 but has good propagation for user 2).

- Sub-carriers can be coordinated between neighbouring base-stations.

- OFDM is suited to the use of smart antennas because different sub-carriers can be allocated and manipulated for different users (for Mobile WiMAX).

- OFDM is modern, hip and not old fashioned and fusty like CDMA.

All of this adds up to claims of much superior performance over CDMA. However, readers need to be slightly cautious. Firstly, CDMA is being enhanced with features such as equalisers that reduce same-cell interference. Secondly, mobility – moving users and passing buses cause the sub-carriers of

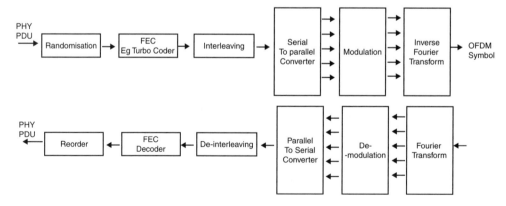

Figure 5-7. Generation of OFDM symbols. (Source: Author. Reproduced by permission of © Dave Wisely, British Telecommunications plc.)

OFDM to shift frequency (a nonlinear operation!) and this reduces their orthogonally. Finally WiMAX requires that the power amplifier runs at well below its optimum region for power efficiency. This comes about because it is possible for all the OFDM signals to peak together – so the amplifier must be run such that this still falls in the amplifier's linear region. Technically it has a high peak to average power ratio. Power efficiencies as low as 20% for the power amplifier are often quoted (cf. 70% for the GSM power amp say). We will look into some of these issues in more depth as we explore the most exciting instantiation of a mobile OFDM system – WiMAX.

5.4 WiMAX MAC Layer

Figure 5.8 shows the components that WiMAX uses in the MAC and Link layers – as with WLANs higher layer functions are not specified in the IEEE standards. The WiMAX forum has, however, created an architecture for the higher layer functions – such as service creation with an IMS – but this is not part of the 802.16 specifications and will be described later. Here we concentrate on the MAC and PHY layers (a bit tedious I know but that is where the WiMAX "magic bullets" are based and, having ploughed through a load of stuff on OFDM you are half way there already – have a d and b[9] and come back)

The MAC layer is divided into a convergence Sub Layer (CS layer) and a common part Sub Layer (CPS layer) – which is responsible for putting the ATM cells, IP packets, Ethernet frames or Sanskrit characters into a common form, header suppression and QoS classification (Figure 5.8). VoIP is a good example of a service needing all these: VoIP can benefit from header suppression as the packets have a small payload and the headers are fixed and need a QoS class with low latency suitable for either constant bit rate or variable rate (VoIP with silence suppression). The CS layer provides QoS by mapping the VoIP packets onto a service flow (each of which is associated with one of five QoS classes in Mobile WiMAX). The mapping could be made in a number of ways; for example the VoIP packets might be marked with DiffServ code points or there may be signalling from an IMS client on the terminal direct to the MAC layer.

The MAC layer (Figure 5.7) also has the following functions:

- Set up and manage the service flows to and from the base station (admission control);

- Schedule the traffic to meet the QoS requirements (in both up and downlinks);

- Perform handovers to neighbouring cells;

- Encrypt the data;

---

[9] Drink and biscuit silly – "oh daddy this museum/chapter is very boring – can we have a d&b" – small girl aged 4–11.

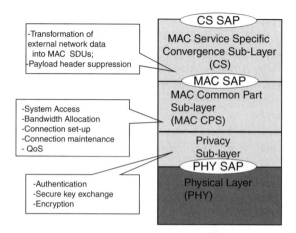

-Transformation of external network data into MAC SDUs;
-Payload header suppression

-System Access
-Bandwidth Allocation
-Connection set-up
-Connection maintenance
- QoS

-Authentication
-Secure key exchange
-Encryption

CS SAP

MAC Service Specific Convergence Sub-Layer (CS)

MAC SAP

MAC Common Part Sub-layer (MAC CPS)

Privacy Sub-layer

PHY SAP

Physical Layer (PHY)

Figure 5-8. WiMAX MAC and link layers. (*Source:* Author. Reproduced by permission of © Dave Wisely, British Telecommunications plc.)

- Recover errors using backward error correction (ARQ);
- Power management (sleep and idle modes).

In this section we will consider each in turn – if only to see that WiMAX is nothing like WLAN but really a fully fledged cellular system.

5.4.1 Downlink QoS and Scheduling

The set up and management of flows is essentially an admission control problem – describing what is wanted (bandwidth, type of traffic or QoS class) and signalling this to the base-station which is in complete control of the resources (i.e. uplink and downlink bandwidth) in the cell. Compare this to a WLAN where the base-station and the terminal "compete" for resources: WiMAX has a much more complicated MAC, in many ways similar to that of 3G. Once a service flow is accepted then the complex scheduler is responsible for allocating transmission opportunities. Figure 5.9 shows an outline of the downlink scheduler. The actual scheduling algorithm is proprietary but likely to be based on well known schemes such as weighted fair queuing or round robin.

Each connection or session between a base-station and a terminal is known as a service flow and has a Service Flow Identifier (SFID) which is allocated to a QoS class (of the five available for Mobile WiMAX) and has a parameter set associated with that flow (e.g. maximum data rate, average data rate etc.). The service flow can be provisioned (by a management plane, say) or admitted on request. Admission is a two stage process – once admitted, such a flow only becomes active after the resources are fully committed following end-to-end negotiation. Active and provisioned flows have a MAC Connection Identifier

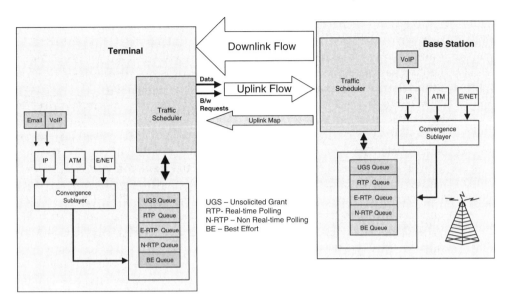

Figure 5-9. Downlink QoS scheduling (QoS classes are explained later). (*Source:* Author. Reproduced by permission of © Dave Wisely, British Telecommunications plc.)

(CID) associated with them (confusingly, admitted but not yet active flows lack a CID!). There are a number of CIDs reserved for system use – e.g. 0X000 is used for initial ranging and 0XFFFB is used for sleep mode multicast – just to give you a flavour of this important system knowledge. Basic, Primary and Secondary management connections are established between the terminal and the BS when the terminal connects to allow the exchange of management and signalling messages. The basic connection is used to exchange short, time critical, messages whereas the primary connection covers management messages for:

• Ranging;

• Access requests;

• ARQ feedback;

• Mesh support;

• Smart antenna support;

• Handover;

• Power mode and sleep support.

The secondary connection is not concerned with the MAC but with Layer 3 support messages – such as DHCP and SNMP. Thus a terminal will have (usually) three management connections in the control plane layer as well as a number of transport connections in the data plane layer that are used to transport the actual data flows. The management CIDs are set up at initial ranging – as the terminal latches on to a new BS (or signs-on to the network). The same CID applies to up and downlink legs of the connection.

In terms of QoS each CID has only one QoS class and its own set of parameters that are used by the scheduler. Packets are classified (e.g. based on type (HTTP or FTP, say) or port number or DiffServ marking) to an appropriate CID for transport. Each new application (VoIP, video streaming, WWW and email) is likely to invoke a new transport connection. The QoS classes are described in a further section – in conjunction with uplink scheduling – to which they are inextricably linked. The set up of the CIDs and QoS classes is likely to be part of the client that will run in association with the WiMAX card or dongle. HSPA cards today often come with clients that perform network selection, and http page compression today and have the capability to offer different levels of QoS. Users bypassing the client are likely only to be offered a fixed QoS class as QoS differentiation is considered a premium service.

5.4.2 Handover

The next MAC function is handover –these are of three types:

- Hard handover – break before make;
- Fast BS Switching (FBBS) – where the terminal does not complete full procedures at the new BS;
- Macro Diversity Hand Over (MDHO) – where the terminal communicates with a number of base-stations.

The actual handover process is not specified in the standards and exists only as a framework within the WiMAX forum. The essential steps of this framework are:

- Cell reselection – the terminal either receives information about neighbour candidates from its own BS or it negotiates scanning intervals to seek out candidate BSs.
- Handover decision and initiation – the decision can be made either in the terminal or by the network.
- Synchronisation to a target BS – this includes initial ranging (i.e. what is the time delay to the BS?).
- Ranging and network re-entry – this can include: handover ranging, authentication and key establishment and IP connectivity. This step can be shortened by the old BS sending information direct to the new BS.

For fast (FBSS) and diversity (MDHO) handover – both of which are optional in the standards –the base-stations and frame transmissions must be synchronised[10] (to better than the cyclic prefix – essentially the delay spread) and BSs can share MAC layer contexts between themselves. In fast handover the terminal only communicates with a single BS (called the anchor BS) but maintains synchronisation with the diversity set of BSs that in turn keep information on the terminal. Fast handover can then be achieved – very quickly and with a minimum of signalling – by changing the BS designated as the anchor BS. In diversity handover, however, the terminal transmits to and receives from multiple BSs. Diversity combining (i.e. all the frames from the different BSs are combined using a clever algorithm to give one with lower errors than any of them) is used at the terminal and diversity selection (the frame with the lowest number of errors is picked) at the BSs receiving the frames. Diversity handover, amounts to adding or dropping BSs from the diversity set depending on the signal quality from each. The mobile WiMAX profile requires only hard handover to be supported and the target is for sub 50ms handover to allow VoIP and real-time video to be supported. In actual fact it is likely that "wave 1" mobile products – available in 2008/9 – will have only proprietary fast handover techniques from day 1 as network level support for handover is still being finalised in the WiMAX forum.

5.4.3 Security in WiMAX

The MAC layer also provides authentication and encryption functionality. In WiMAX each terminal has a unique X.509 certificate –the normal sort you get on shopping and banking websites – issued by a Certification Authority (e.g. VeriSign) and installed by the manufacturer. The certificate contains the terminal's public key and 48-bit MAC address – the terminal also has a matching private key in the usual public/private key arrangement.[11] There are two generations of security – PKMv1 (Privacy Key Management) – standardised in 802.16-2004 and PKMv2 with mobility additions. PKMv1 allows the terminal to send an authorisation request to the BS containing its certificate – the base-station is then able to verify the certificate is valid (i.e. the terminal is who he/she/what it claims to be) and can then determine if they are authorised

---

[10] BS synchronisation is generally done with GPS receivers at each BS. At the physical layer all BS need to be synchronised as without synchronisation then there would be the possibility of one BS in the uplink part of the frame whilst another is transmitting in downlink. This would generate an unacceptable amount of interference for the uplink.

[11] Anything encrypted by the public key (which is given out freely) can only be decrypted with the private key and, conversely, anything encrypted with the private key can only be decrypted with the public key. Certificates allow entities (like websites) to prove who they are by evoking certificate authorities who issued them in the first place. Each certificate has the owner's name, expiry date and name and signature of the issuer. The public key of the issuer can then be used to establish the validity of the certificate – in the case of a website it confirms that the certificate was issued to (say) www.trolleydolleys.com. This authenticates the site and my browser and the site server then establish keys to encrypt the details of my credit card.

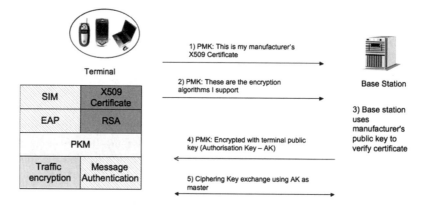

Figure 5-10. Security in WiMAX. (Source: Author. Reproduced by permission of ©
Dave Wisely, British Telecommunications plc.)

to access the network. The base-station's response to the authentication request
is either rejection or success – in which case an Authorisation Key (AK) is sent
(this is encrypted with the terminal's public key – which can only be unlocked
with the private key held in the terminal). The AK is used to derive and exchange
a number of keys with the BS including session keys for traffic encryption and
message authentication. PKMv2 adds the need for two-way authentication
(i.e. the base-stations prove they are genuine) and mandates AES encryption
(a high strength encryption algorithm). In addition PKMv2 adds support for
Extensible Authentication Protocol (EAP) – a flexible protocol (described in
Chapter 3 in connection with WLANs) that can use, for example, SIM cards.
Thus WiMAX terminals could have 2G or even 3G SIM cards. The overall
security features of WiMAX are illustrated in Figure 5.10. In practice the
terminal certificate is unlikely to be sufficient for authentication of the user –
since it is tied to the terminal. Operators are likely to add SIM cards or other
security mechanisms that are tied to the user and that enable handover to 3G
cellular-based technologies. WiMAX standards support this through EAP
(Extensible Authentication Protocol) either instead of or in addition to the
X509 terminal certificate – allowing both the terminal and the user to be
authenticated.

5.4.4 ARQ and Forward Error Coding

The MAC layer is also the home of an advanced ARQ scheme – with options for
a bewildering range of ACK possibilities (cumulative, selective, selective with
cumulative and so on). ACKs are also piggybacked in the MAC frame headers to
save time and bandwidth. When the OFDMA (see later) PHY layer is in
operation (in practice this is Mobile WiMAX) a Hybrid ARQ – very similar
to the Hybrid ARQ described in Chapter 4 in respect of HSDPA – is used. There
are several modes of the HARQ including incremental redundancy – where up
to four "versions" of a frame are sent sequentially in an attempt to allow the

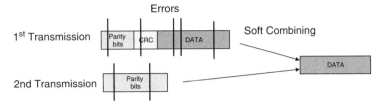

Figure 5-11. HARQ with incremental redundancy – as used in WiMAX. (Source: Author. Reproduced by permission of © Dave Wisely, British Telecommunications plc.)

receiver to recreate the original frame. A second mode is Chase Combining – which simply means repeating the original frame. One of the features of WiMAX is its low latency – in the low 10s of milliseconds and the error coding has been especially designed to support this from day 1 – as opposed to "bolted on" to an inherently long latency system like UMTS. Figure 5.11 shows the incremental redundancy mode of the HARQ.

5.4.5 Idle Mode and Power Saving

The introduction of a sleep mode in WiMAX potentially offers considerable power-saving advantages. When the terminal enters the sleep mode (Figure 5.12) it does not receive or transmit frames for a set sleep interval – after this it enters a listening interval during which the BS will inform it of any waiting traffic. The size of the sleep intervals is flexible but would need to be every few seconds for any terminal expecting incoming VoIP calls, say. The sleep interval might also increase as the battery on the terminal runs down – trading responsivity for extended battery life.

WiMAX also supports an idle mode in terminals – which is used in conjunction with paging. Idle mode differs from sleep (and can be used in conjunction

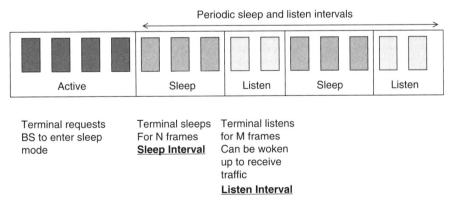

Figure 5-12. Sleep mode in WiMAX. (Source: Author. Reproduced by permission of © Dave Wisely, British Telecommunications plc.)

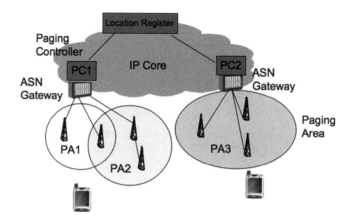

Figure 5-13. Paging in WiMAX. (Source: Author. Reproduced by permission of © Dave Wisely, British Telecommunications plc.)

with sleep). In idle mode the terminal is able to receive broadcast information from BSs without registering them – so that it can simply monitor the broadcast CID without ranging, handover or other (power hungry) activities. BSs are divided into paging groups and broadcast paging messages for any terminal that has an incoming session request (e.g. a VoIP call). Clearly the terminal must be in a listen interval if it is using sleep mode as well and this is taken into account by the BSs. When a terminal receives a paging request it then performs ranging and sets up management CIDs followed by a transport CID for the VoIP or incoming session. If the terminal moves out of the current paging area it needs to re-register so that the macro mobility management system (in WiMAX it is Mobile IP) – can direct paging requests to the correct paging area – Figure 5.13.

This is very different from WLANs – where there is no idle mode or paging concept and power consumption is about a factor of 10 times higher than in typical 2G or 3G devices. It is interesting to see that the IEEE 802.11v working group is adding idle mode and paging to WLANs to reduce power consumption in mobile devices (Chapter 3). It also illustrates that WiMAX is not really "WLAN on steroids" – it is a fully fledged cellular system – 10 times as complicated and much more sophisticated.

5.5 WiMAX PHY Layer

OFDM is one of the two "magic" bullets of mobile WiMAX – the other is smart antennas – which are described in a later section. As the watchful reader will have noted there is nothing particularly novel about the WiMAX MAC layer – it performs many of the functions that an HSDPA or 3G MAC layer does. True it has all the functionality that 802.11 WLANs are trying to "bolt on" – such as QoS, power saving, security etc. – but nothing that cellular doesn't already have. No, it is the use of OFDM in a flexible MAC scheme, together with the use

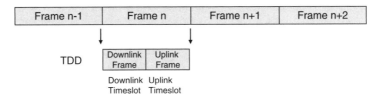

Figure 5-14. TDD operation in WiMAX. (Source: Author. Reproduced by permission of © Dave Wisely, British Telecommunications plc.)

of advanced antennas that is the secret to WiMAX differentiating itself, in terms of performance, from HSDPA.

Mobile WiMAX currently is only specified as a TDD system – in other words up and down links share the same frequency, alternating up and down transmissions. In WLANs the uplink and downlink transmissions follow a rather haphazard sequence – as the base-station has to compete with the terminals for the right to transmit. In WiMAX the base-station has absolute control with transmissions divided into frames, each of which has both a downlink and an uplink subframe (Figure 5.14). This arrangement has the advantages that it enables a cheaper transmitter/receiver as it: reduces the component count, is well suited to smart antennas[12] and it allows the base-station to alter the up/down capacity split as traffic changes. The latter point is important as DSL traffic – considered to be a possible pointer to future WiMAX traffic – is highly asymmetric (typical 4 or 5:1). Figure 5.15 shows more detail of

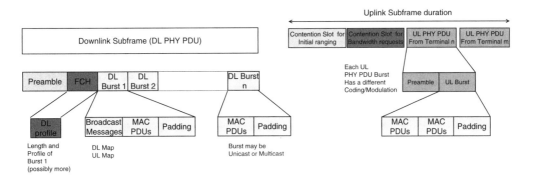

Figure 5-15. WiMAX Downlink (left) and uplink (right) subframes. (Source: Author. Reproduced by permission of © Dave Wisely, British Telecommunications plc.)

---

[12] The uplink radio propagation (dealy spread, signal level, user spatial location etc) is used to estimate the downlink channel propagation which is required to set the antenna weights. If the uplink and downlink frequencies are the same then the estimate is very accurate – but this falls away as the uplink and downlink frequencies diverge.

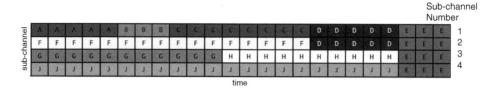

Figure 5-16. WiMAX Channelisation (Courtesy of BTTJ) four sub-channels shared by nine (labelled A to J) users. (Source: BTTJ. Reproduced by permission of © British Telecommunications plc.)

the WiMAX frame structure – we will come back to many of the features shown here in the next few sections.

Mobile WiMAX uses OFDMA – Orthogonal Frequency Division Multiple Access for both up and downlinks. OFDMA is a flexible multiplexing scheme in that it can share OFDM carriers between users dynamically on a very short time scale. In Mobile WiMAX there are 512 or 1024 sub carriers and these are divided into groups – called sub-channels. The actual distribution of the active sub-carriers making up a sub-channel can be selected in a number of ways. They can be contiguous – which might be used to avoid other sub-channels in a neighbouring cell – or distributed – which would provide better resilience against interference or fast fading, say. Users are allocated one or more sub-channels in a given time slot (Figure 5.16) – allowing a very flexible, two-dimensional (time and sub-carrier) allocation scheme. In Figure 5.16 it is easy to see the flexibility of this scheme. User J might have particularly good radio transmission over sub-channel 4, for example, and user E might be a high priority user requiring a very high throughput. Sub channels might also have lower interference in different directions – as we will describe later –user G might be east of a base-station where neighbouring cells are not using sub-channel 3, say.

Without the competition for access slots used in WLANs – the base-station needs to inform the terminals of the times when they should be listening and when they should transmit. This is achieved by the base-station publishing downlink and uplink maps at the start of each frame (Figure 5.15) – detailing exactly when and to which sub-channels each terminal such listen or transmit – the maps are shown in detail in Figure 5.17. Clearly for this to work it is necessary for each terminal to be accurately ranged – i.e. know how long it takes for its signal to reach the base-station (radio travels a ns per foot[13] – so one mile is 5micro sec or so – at 10Mbit/s that is 50bits per mile!)

The actual map can get quite complicated and these are re-calculated for each frame – the frame length can be between 2 and 20 ms – as determined by the BS but in Mobile WiMAX 5ms is the only mandated value. There are two further complications with the map that need to be mentioned. Firstly, the

---

[13] Or for those readers who use the metric system – just substitute 30 cm for a foot, 1 m for a yard, 1.8 km for a mile and 201 m for a furlong. Rods, perches and ells are starting to go out of use as measures in England and so I won't refer to them in the book unless absolutely necessary.

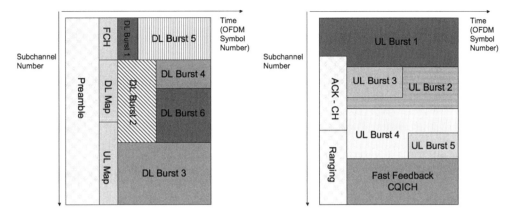

Figure 5-17. WiMAX Transmit maps (Downlink left and uplink right). (Source: Author. Reproduced by permission of © Dave Wisely, British Telecommunications plc.)

downlink subframe begins with a long preamble[14] – to allow listening terminals to obtain synchronisation – and this is followed by the Frame Control Header burst containing information on at least the following burst profile (length, sub-channels, coding etc.). The FCH is coded using the most robust coding available (BPSK ½ rate). It is actually the burst following the FCH that contains the downlink and uplink maps as well as other broadcast messages.

The use of OFDMA/TDD and the per-frame evaluation of maps makes WiMAX very flexible. It is possible to swap sub-channels as users move around, synchronise with neighbouring base-stations so that two sub-channels are never used at same time in the same place and for the uplink/downlink balance to be altered in real-time. It is also possible with certain adaptive antenna techniques to beam form to individual users on a sub-channel basis.

5.6 WiMAX QoS

The astute (awake?) reader will notice that the above description of the MAC avoids talking about uplink QoS. Uplink QoS in WiMAX is complicated (it took me days to figure it out – don't even think about reading the standard unless you are suffering sleepless nights!). Uplink QoS in fact mixes a lot of functions that need to be separated in order to understand them. It also uses confusing names that imply functions do one thing but actually do another! Personally I think that this section of the book is worth the cover price alone!.).

Firstly, there clearly needs to be a mechanism for the terminals to signal their bandwidth requirements – these can, of course be incredibly varied from a VoIP

---

[14] The preamble is often transmitted at a higher power level than the data sub-carriers, typically 2.5 dB higher.

call with fixed size packet requiring low latency and produced in a continual stream to WWW browsing that generates bursts of activity with medium latency requirements but only periodic activity. Once these requirements are signalled to the BS it then needs to run an uplink admission control system, operate an uplink scheduler and communicate the output of the scheduler to the terminals in the uplink map. Terminals then need a way of feeding back information on any backlogs (the error rate might be higher than expected) or any changes in circumstances.

It is useful to distinguish four separate components of the uplink QoS scheme:

- Mechanisms for signalling uplink requirements;
- Five QoS classes for mobile WiMAX (four for fixed WiMAX);
- Mechanisms for granting uplink bandwidth;
- Scheduler that runs in the base-station to perform admission control and manage real-time uplink bandwidth allocation.

The mechanisms for signalling are quite diverse – apart from the UGS/ertPS (Unsolicited Grant Service and extended real-time Polling Service – see below) – each request is for an allocation of bandwidth within a single frame, i.e. the terminal has to signal just to get an allocation in the next frame.

The first signalling mechanism is that the base-station polls (i.e. contacts one at a time in sequence) the terminals providing them with either unique or contended opportunities to request bandwidth. The second method that terminals can use to signal bandwidth requirements is piggybacking them with data – i.e. new requests can be included within the headers of data frames. In addition the actual allocation of bandwidth is to the terminal as a whole – not at a specific connection (CID). So the terminal is able to allocate this bandwidth between applications as it sees fit. Moreover it is able to use this bandwidth for data or requests (or even requests piggybacked in data) as is needed. This is called bandwidth stealing but is restricted in certain QoS classes only – as will be explained below.

The five QoS classes are shown in Table 5.3. The actual names are confusing and refer to the main mechanisms of actually allocating bandwidth request opportunities. The UGS (Unsolicited Grant Service) is aimed at real-time data streams with fixed-sized packets at regular intervals – such as VoIP (without silence suppression) or E1 TDM circuits, say. When a UGS request is admitted by the base-station it then allocates regular uplink bandwidth for the data – no new requests are necessary from frame to frame to keep the bandwidth flow moving (Figure 5.18). Terminals are permitted to piggyback requests or to steal this bandwidth to make further requests – which might be for other QoS class traffic say. The ertPS (extended real-time Polling Service) is similar to the UGS save for the fact that uplink allocations are not fixed in size but dynamic (i.e. the terminal uses piggybacking or bandwidth stealing to indicate the size of the bandwidth needed in the next slot). ertPS was added for

Service types	Description
Unsolicited grant service (UGS)	UGS is designed to support real-time data streams, consisting of fixed-size data packets issued at periodic intervals, such as backhaul and voice over IP.
Real-time polling service (rtPS)	rtPS is designed to support real-time data streams consisting of variable-sized data packets that are issued at periodic intervals, such as MPEG video.
Non-real-time polling service (nrtPS)	nrtPS is designed to support delay-tolerant data streams consisting of variable-sized data packets for which a minimum data rate is required, such as FTP.
Extended Real-time Polling Service (ertPS)	Support real-time service flows such as Voice over IP services with silence suppression. Terminal can request to change the size of the request opportunity
Best effort (BE)	BE service is designed to support data streams for which no minimum service level is required and which can be handled on a space-available basis.

Table 5-3. WiMAX QoS classes.

802.16e and is intended for VoIP with silence suppression and some video applications.

The rtPS (real-time Polling Service) is designed for video streams where the data packets vary in size but need to be transported at regular intervals. This service provides (through unicast polling) regular opportunities for the terminal to request bandwidth.

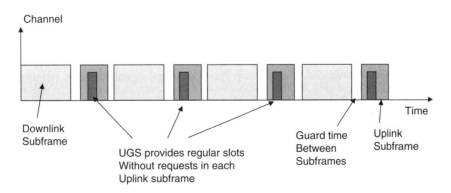

Figure 5-18. WiMAX UGS QoS class operation. (*Source:* Author. Reproduced by permission of © Dave Wisely, British Telecommunications plc.)

Scheduling service	Piggyback grant request	Bandwidth stealing	Unicast polling	Multicast polling
UGS	Yes	Yes	Poll-Me bit can be used	No
ertPS	Yes - extended	Yes	Yes	Yes
rtPS	No	No	Yes	No
nrtPS	No	No	Yes	Yes
BE	No	No	Yes	Yes

Table 5-4. Request and grant mechanisms for the WiMAX QoS classes.

Next is the nrtPS (nonreal-time Polling Service). This is designed for services like FTP or web browsing – where a degree of delay tolerance is permissible. In this case the base-station can offer (depending on its scheduler algorithm) either unicast polling opportunities – meaning that the terminal being polled gets unique access to the opportunity to request bandwidth – or multicast polling – meaning that the base-station offers a group of terminals a chance to compete for the right to request bandwidth. The contention slots follow the ranging slots at the start of the uplink frame (see Figure 5.15). Basically each terminal has an initial window and chooses a number of back-off slots within that window – if a collision with another station occurs (detected by the fact that the request is not answered) the terminal increases the window size and tries again – very like the WLAN MAC operation.

The final class – Best Effort (BE) can be serviced by the base-station with unicast or contended request opportunities as the base-station scheduling algorithm sees fit. Overall the standard allows great flexibility of how bandwidth is requested and allocated and so, in practice, the actual performance will mostly come down to how well the admission, congestion and scheduling algorithms in the base-station work. Table 5.4 attempts to show how the request/grant mechanisms can be mixed and matched for each QoS class.

5.7 Spectrum

It is expected that WiMAX will be deployed in only a limited number of bands which include:

Licensed bands:

2.3GHz – including the USA, Korea (WiBro uses 2.3GHz)

2.5GHz – UMTS extension band (Europe), USA, SE Asia and Central/S. America

3.3GHz – SE Asia

3.5GHz – Europe (France and UK have allocated licences), Central/S. America

Unlicensed bands:

5.8GHz – U- NNI band

The actual rules in each band vary between different countries: for example, the 3.5GHz licences offered in the UK in 2003 prohibited mobile services to be offered but they have now been relaxed (despite opposition from mobile operators).

Mobile WiMAX would only use licensed bands and the WiMAX forum have produced mobile WiMAX profiles (Table 5.2) that support 2.3, 2.5, 3.3 and 3.5GHz with channel sizes between 1.5MHz and 20MHz. This is an important feature of WiMAX – allowing it to operate in small segments of spectrum that might become available, although this greatly reduces the efficiency of the system.

The frequency band used also affects the spacing of base-stations. As a general rule, the cell radius is proportional to the carrier frequency used. Thus a WiMAX network at 3.5GHz could require roughly 60–80% more base-stations than a 2.1GHz 3G network – in reality it is expected that for full in-building coverage and good mobility support Mobile WiMAX will have to use frequencies of 2.6GHz or lower. The 3.5 and 3.3GHz bands are more likely to be used for out-building coverage (with internal repeaters – possibly using WiFi) where mobility is restricted to nomadic use. In other words campuses and city centres where users were sitting at café tables or in parks.

Fixed WiMAX systems have been operating since 2005 – with many countries having one or more systems operational or in trial. Examples of recent news on fixed WiMAX are (see the references at the end of the chapter for sources of news):

- Sri Lanka – Dialog Telekom, has introduced broadband over WiMAX technology in 55 main cities and 75 towns with speeds up to 4 Mbit/s (Sri Lanka has only 160 000 Internet connections for a country of 21 M).

- Freedom4 – formally Pipex is set to launch a new WiMAX market in the UK in Q1 2008 operating in licensed 3.5GHz spectrum, offering a fixed WiMAX service in the city of Manchester with plans to extend to 50 cities in the UK.

- Bollore Telecom – one of 12 WiMAX licence holders in France is planning a pilot of nomadic services – based on 802.11e Motorola equipment, in the 3.5GHz band in Paris.

- India will have up to 21 million WiMAX subscribers by 2014, according to a report co-authored by Maravedis and Indian market research firm Tonse Telecom.

- Umniah, Jordan's newest mobile communications provider has launched WiMAX services under the UMAX brand with services replacing DSL and cable services with a wireless alternative with speeds of up to 2 Mbps, with a range of up to 15 km.

Mobile WiMAX is currently operating in Korea in the form of WiBro from June 2006 – which really was developed before WiMAX and later incorporated within the developing standard. There are currently two operators of WiBro in Korea KT and SKT. Both systems work in the 2.3 GHz band but don't include smart antenna technology. KT has established coverage of Seoul and 17 universities with 80 cities due for coverage in 2008 and nationwide coverage expected by 2009. In 2007 KT had 70 000 USB modem users and 12 600 phone users for WiMAX-SKT much less. Plans are very aggressive, however, with 1.2M by 2009 and 4M users expected by 2012. Currently WiBro is cheaper than HSDPA in Korea ($12 (US) per month for 1 Gbyte or $24 for unlimited data compared to $30 for 2 Gbyte for HSDPA). Hong Kong's Office of the Telecommunications Authority (OFTA) will auction two licences in late 2008 in the 2.3 and 2.5 GHz bands. In Japan the decision as to which two consortia will be awarded a mobile WiMAX licence (2.5 GHz) is currently under consideration. KDDI has been running a trial since 2005 which has shown handover with EVDO systems, smart antennas and convergence over several access technologies.

In Europe most interest centres on the so-called "UMTS extension band" (2500 MHz to 2690 MHz) – which has been declared 'technology neutral' and is currently being auctioned around Europe. After this, the next major opportunity for spectrum will be when analogue TV transmissions cease in (2010–12 in the UK). This lower frequency spectrum (400–700 MHz) is particularly appealing as this would allow network coverage to be quickly established with fewer base-stations (due to the increased propagation range) – greatly reducing start-up costs. Larger cells would reduce the bandwidth density for capacity limited cells. Further exploratory talks are taking place in the UK about re-farming the 900 MHz GSM spectrum – some of this may (Ref. 6) be offered in the future. Regulations on the use of the unlicenced 5 GHz band should be harmonised across the EU shortly, following a decision by the European Commission. All member states will have to comply with the EU regulations on the licence-exempt use of the 5150–5350 MHz and 5470–5725 MHz frequency bands.

In North America Sprint Nextel have just began to test WiMAX systems with their own employees (Q1 08) with a commercial rollout of Q2 planned (Ref. 14). There is also speculation that AT&T – with many 2.3 GHz licences – is preparing for trials or even launch of a WiMAX in 2008.

5.8 Smart Antennas

Apart from using OFDM the other "magic bullet" of WiMAX is smart antennas. In GSM and 3G there are two types of antenna: In rural areas a single antenna radiates omni-directionally and in urban areas there are three separate antennas that transmit into different sectors. These are not smart antennas – because they are really three cells of a standard cellular system just sharing a common (and expensive) base station and back haul. The other antenna technique that

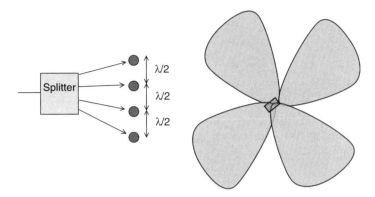

Figure 5-19. Four element transmitter and radiation pattern. (Source: Author. Reproduced by permission of © Dave Wisely, British Telecommunications plc.)

was introduced earlier – diversity – is not a "smart" technique either. In diversity two or more antennas transmit/receive the same signal but are separated (either by distance or polarisation) so that the signals from each are uncorrelated. They are typically combined with weights related to the signal to noise/interference on each branch.

To see how smart antennas can (in theory) dramatically improve spectral efficiency imagine a transmitter with four independent transmitting elements (Figure 5.19). If each antenna radiates in an omni directional pattern using the same frequency then they will interfere with each other – if there is no coordination between the antennas then the result will be heavy interference. If, however, the antennas are spaced at a regular "short" interval (so that the radio paths for each are the same – say ½ wavelength – 7.5 cm at 2 GHz) and that the same signal is fed to each antenna then at some locations the various signals interfere coherently to produce a strong signal and at others they interfer destructively and the signal falls to nothing. This is, of course, nothing more than good old diffraction – the row of antennas could be a row of slits being illuminated by coherent light – or the bridge piers across a river producing a wake pattern. The number of "lobes" (fringes) is equal to the number of elements in the array – six elements = six lobes. The lobes can also be steered around by varying the phase of the signals into each antenna and, even more importantly, the power in each lobe can also be controlled by adjusting the phase and amplitude of each element of the array – just as the blaze (the shape of the groove) of a diffraction grating determines the power in each order.

This use of smart antennas is called beam-forming (and sometimes Adaptive Antenna Systems – AAS) – Figure 5.20 shows how it can be used to increase the signal to noise/interference ratio (and hence data rate) for distant users – without increasing the overall output power and, when used in a receiver (as opposed to a transmitter array) to block sources of interference, to increase sensitivity in the direction of terminals.

In fact there is a range of beam-forming techniques that can be used – including spatial multiplexing: sending different signals to different users who

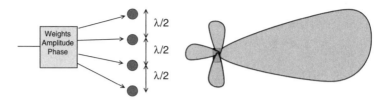

Figure 5-20. Beam forming in WiMAX using phase and amplitude adjustment for a four element transmitter. (*Source:* Author. Reproduced by permission of © Dave Wisely, British Telecommunications plc.)

are spatially separated using the same frequency/code. This could be achieved for users – say North and South of a base-station – by having a beam that has a null south and a strong lobe north – modulate this with North's data. Then calculate the phase and amplitude needed to do the opposite and modulate this with South's data. Although these could be separate transmitter arrays it's also possible to use just one array, since the response of an array is linear then you can just combine the two signals on the one array to get the same effect. Of course the downside is that each only gets half the power. Combine this with a receiver array that can simultaneously receive data from North and South and eliminates strong interferers east and west you might (in this, admittedly, contrived situation) have increased the capacity several times over the case of a three sector antenna (assuming a constant transmit power).

802.16e offers three advantages for smart antennas. Firstly, it is a TDD system – meaning the up and downlink use the same frequency band – and so the signals experience exactly the same radio conditions. The weights for the downlink transmitter can be set based on the received uplink signals (although this becomes less true as the speed of movement increases). In an FDD system (like 3G) the use of smart antennas requires complicated estimates of the downlink channel being reported back from the terminals to the base-station. The second advantage that 802.16e offers is that beam forming is most effective in a narrow band system – remember that when you did interference experiments with light you needed either a laser or a lamp with a very narrow optical filter – the light needed to have a narrow spectral range to form any fringes. Much the same is true of radio systems – a wideband technology (like CDMA) will not form tight lobes in the way that each OFDM carrier will. In OFDMA beam forming is done on a per sub-channel basis – so one sub-channel can have a null where that sub-channel is in use by a neighbouring base-station (Figure 5.21 – which shows the standard three sector cells arrangement – and Figure 5.22 which shows WiMAX with a frequency re-use of 1 using beam steering). In fact it is unlikely that WiMAX would be able to achieve a frequency re-use factor of 1 without the interference reduction of beam forming. The first wave of WiMAX products will have a three frequency re-use pattern – meaning that three frequencies will be needed to provide separation between cells (although one supplier has a four sector system with two or four frequency reuse).

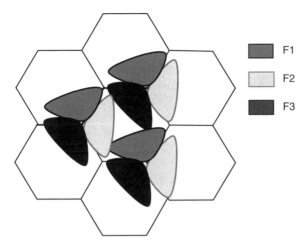

Figure 5-21. Normal three-sector antenna pattern. (*Source:* Author. Reproduced by permission of © Dave Wisely, British Telecommunications plc.)

The 802.16-2004 and 802.16e standards include support for beam forming – although it is not included in any fixed profiles it is mandated to be supported by the terminals and optional for the base-stations of Mobile WiMAX. Support for beam forming includes dedicating a specific signalling zone for terminals to transmit in – allowing the base-station to measure the interference on different

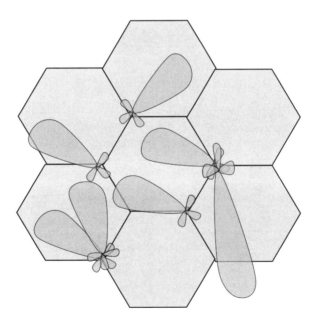

Figure 5-22. WiMAX beam forming to reduce interference. (*Source:* Author. Reproduced by permission of © Dave Wisely, British Telecommunications plc.)

sub-channels. This is added to the MAC/PHY specification and makes adding adaptive antennas much easier than retrofitting them to CDMA systems, say.

The most powerful form of smart antennas is something called MIMO – Multiple In Multiple Out. MIMO is considered more effective at the moment for throughput gains than beam forming. So far we have only talked about the base-station having a transmitter/receiver array – the terminal has just had a single receiver/transmitter – mostly because more antennas mean more processing and power consumption and because of space limitations (at 3 GHz a wavelength is 10 cm and even a quarter wavelength spacing requires 2.5 cm between antennas). In MIMO both ends use array technology (Figure 5.23) and, in a near miraculous way, for an N-element transmitter array and an N-element receiver array, the capacity between the two ends can be up to

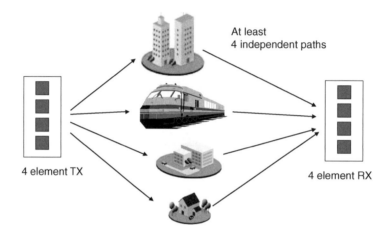

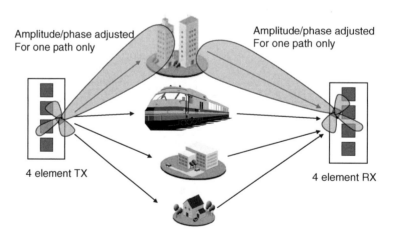

Figure 5-23. MIMO set up. (*Source:* Author. Reproduced by permission of © Dave Wisely, British Telecommunications plc.)

N-times greater than that of a single element system. In fact it only works where there is extreme diversity in the radio environment and it is easy to see, in a hand-waving way, how this comes about. If you are not convinced and want to see the maths try Ref. 7 but have the aspirins handy. Imagine the radio path between the transmitter and receiver arrays of Figure 5.23 is dominated by four main "rays" – one that bounces off a skyscraper, one that bounces down the sides of the streets, one that diffracts round the trees in the park and one that scatters from the prison railings! If you could "see" radio energy instead of optical energy you could look from the receiver end towards the transmitter and you would see four bright spots – they would be coming from different directions in space. The trick in MIMO is to then adjust the phase and amplitude weightings of the receiver array such that the array is sensitive to one of these bright spots and has low sensitivity (null lobe) on the other three (Figure 5.23 lower part). Then you calculate the weightings for the other three spots to be "in beam". Conceptually you could have separate receiver arrays – each with the different settings for phase/amplitude – but in practice you can have one array and then apply the different weights to $1/4$ of the signal from each to form four different signals – each with one sensitive to one of the bright spots and not responding to energy in the other spots.

At the transmitter you then do the opposite – each spot at the receiver is caused by rays leaving the transmitter in a certain direction. If you apply appropriate weights to the transmitter array you find that you can select any of the four ray paths from the transmitter and they are resolved independently at the receiver (Figure 5.24). In MIMO the capacity gain comes from sending different data streams along each path simultaneously – in this case giving a theoretical four times capacity increase. The capacity increase has also not necessarily come at the expense of increasing the power– each branch of the

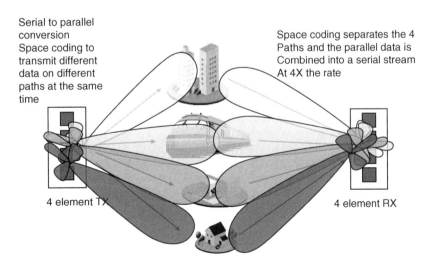

Figure 5-24. MIMO used for capacity multiplexing. (*Source:* Author. Reproduced by permission of © Dave Wisely, British Telecommunications plc.)

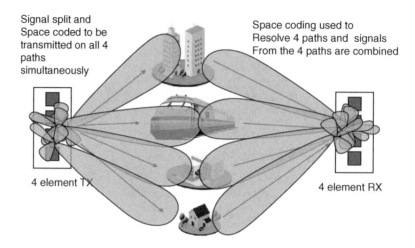

Signal split and
Space coded to be
transmitted on all 4
paths
simultaneously

Space coding used to
Resolve 4 paths and signals
From the 4 paths are combined

4 element TX

4 element RX

Figure 5-25.   MIMO for diversity. (*Source:* Author. Reproduced by permission of © Dave Wisely, British Telecommunications plc.)

MIMO transmission might receive more energy than an omni transmitter and receiver would manage. This is called Spatial Multiplexing. Another MIMO technique is called spatial diversity whereby the same signal is sent on the different paths (Figure 5.25). There are a number of variations of multiplexing and diversity that can be used depending on the precise radio conditions. Where the signal to noise/interference ratio is poor (e.g. at the edge of the cell) then diversity might be more appropriate, whereas, when the paths each have a good signal to noise/interference ratio and can be easily resolved, then multiplexing might be more effective.

Of course nothing is quite that simple – in reality the different paths would have different losses and not all of them would be suitable for the data rate of the best path. Also the "isolation" between the paths would never be complete and that would cause interference between them. In a mobile situation, the paths would also change very rapidly and would need to be estimated and used faster than the rate of change. Finally, for MIMO to work, there need to be many diverse paths between the transmitter and receiver. In a rural environment this is unlikely to be the case and MIMO will have no advantage. Only in dense urban environments is it reckoned to be capable of increasing capacity by a factor of 2 (Ref. 8). Finally, there is the problem of fitting more than two transmitters/receivers in the terminal.

Currently 802.11e provides some support for MIMO and from the options in the standard the WiMAX Forum has selected two schemes (diversity and multiplexing) with two transmit antennas at the base-station and two receive antennas at the terminal. In the uplink the only selected solution is collaborative spatial multiplexing – whereby each terminal has only a single transmit antenna but different users can transmit on the same sub-channels at the same time and are discriminated at the BS because they arrive on different radio paths that can be resolved using appropriate weights on the BS receive antenna elements. This scheme limits the uplink power consumption of the terminals whilst allowing some of the gains of smart antennas.

5.9 WiMAX Network Architecture

The IEEE 802.16 standards only define the MAC and PHY layers – as is the case with WLANs. However, whilst there are developments of WiMAX femtocells that can be plugged into DSL lines there are good reasons why operators will want to enhance this most basic deployment. Not least of which is the need to offer exciting services beyond Inter/Intra net access to recoup the considerable costs of licensing spectrum and building out of what is, in effect, a cellular system. In particular WiMAX networks need to provide:

- Support for handovers across the network.
- QoS support – so that real-time services (VoIP, video etc.) are supported end to end and not just over the air.
- Authentication services (e.g. to validate SIMs, name/password and certificates).
- User profiles and authorisation functions (Is the user authorised for the QoS-level requested?).
- Call/session records and accounting functions.
- Support for VoIP services (since there are no circuits in WiMAX).
- Support for emergency services and lawful intercept.
- Value-added service creation for revenue generation (e.g. portals for music downloads).
- Roaming to other cellular networks.
- Interconnect with the PSTN.
- Support for multicast and broadcast services.

The job of defining the WiMAX architecture has fallen to the WiMAX forum – which has now issued specifications for a network reference model and also interworking guidelines as well as descriptions of the protocols and procedures to be implemented in WiMAX end to end networks (Ref. 9).

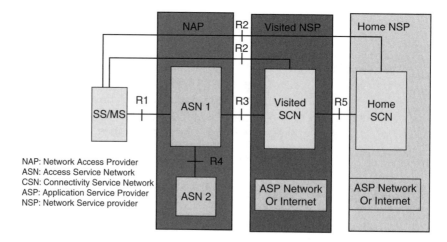

Figure 5-26. WiMAX reference architecture (R6 and R8 are within the ASN). (Source: Author. Reproduced by permission of © Dave Wisely, British Telecommunications plc.)

Figure 5.26 shows the concepts and interfaces of the WiMAX reference architecture. The terminal is called a Mobile Station (MS), the radio network is referred to as an Access Service Network (ASN) and the core network is called the Connectivity Service Node (CSN). Notice that the WiMAX forum have retained the concept of home and visited networks – as is the case in 2G and 3G.

The ASN is the radio network and has one or more ASN gateways (ASN-GW) that are equivalent to RNCs in 3G or BSCs in GSM. The ASN provides a number of defined services:

- Network discovery;

- Location management with the ASN;

- Paging;

- Layer 2 connectivity across the ASN;

- Handover within the ASN;

- Support for IP connectivity functions (AAA, DSCP etc.);

- RRM functions (QoS support, handover etc.).

The ASN will usually have a number of base-stations controlled by one or more ASN-GWs. There are three profiles indicated by the IEEE that specify the split of the above functions between the base-stations and the ASN-GW called (somewhat unimaginatively) A, B and C. Profile B simply leaves the split open to vendors!

Figure 5.27 shows a possible core network design for WiMAX. The key functions of the core network are:

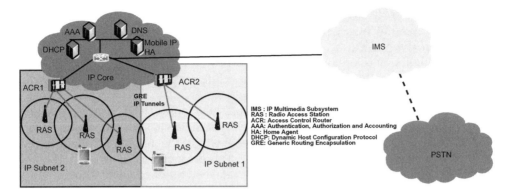

Figure 5-27. WiMAX core network – possible design. (*Source:* Author. Reproduced by permission of © Dave Wisely, British Telecommunications plc.)

- IP connection authorisation and establishment (AAA, IP address allocation);
- IP transport (routing);
- Mobility support (using Mobile IP);
- QoS support;
- User profile and authentication;
- Call/session records and billing;
- Service support (VoIP, location services, downloads etc.);
- Interworking;
- Support for visiting terminals.

Most mobile WiMAX trials to date have used an IMS for service creation and this is the configuration for WiBro systems (if you don't understand the IMS and its architectural elements then have a look at Chapter 6).

5.10 WiMAX Throughput and Performance

Not much in the world of telecoms has created as much controversy – some would say howls of laughter – as when WiMAX was first mooted and figures such as: "WiMAX provides for up to 50 km of service area range, allows users to obtain broadband connectivity without needing direct line of sight with the base station and provides shared data rates of up to 70 Mb/s" – Ref. 10.

"It has a service range of up to 50 km and provides data rates of up to 280 Mb/s per base station": Ref. 11.

The trouble with "headline" figures like these is that GSM could do the same IF you configured it in the right way. My brother-in-law's lorry could do

150 mph if he stripped it back to the cab and fitted a turbo charger etc. – it would go like stink but be pretty useless for his property development business! To actually get 70 Mb/s over 50 km requires fixed, highly directional antennas with line of sight operation, high powers and 20MHz of bandwidth. This is effectively equivalent to a microwave link and, viewed as such, the performance quoted seems much less impressive.

To help the readers with the "sales presentations" of leading WiMAX vendors the key measure that they need to really compare like with like is the spectral efficiency. Spectral efficiency is the overall throughput of a cellular system expressed as the bits/sec throughput per Hz of spectrum used in a cell. There are a number of caveats that need to be attached to spectral efficiency:

- The spectral efficiency of an isolated cell is about two to five times higher than in a cellular system – due to the effects of interference from neighbouring cells.

- The spectral efficiency depends on the traffic type – it is different for circuit voice and burst, Internet-type traffic.

- The spectral efficiency increases (for most technologies) with size of the band being used (e.g. WiMAX in a 20 MHz band is more efficient than WiMAX in a 5 MHz band).

- The bit/s should (but don't always) refer to the useful bits transferred (i.e. excluding re-transmitted or dropped frames and excluding coding bits). This is still 30–50% higher than the rate real applications (such as ftp or www browsing) actually see.

- The spectral efficiency reduces as the mobility (i.e. speed) of the users increases. If most of the users are in cars the efficiency will be significantly lower (for all technologies) than for static users.

- Capacity and coverage are not the same thing. Initially networks will be built to achieve coverage – simply covering the whole city, say (known in the business as the "thin and crispy" option). Later the need to add capacity means that more base-stations are added ("deep pan") and the average distance from terminal to BS reduces – this affects the spectral efficiency of different systems.

With these caveats it is possible to make meaningful comparisons between technologies and see how WiMAX might stack up against HSDPA and 3G, for example. Table 5.5 with figures from the WiMAX forum – Ref. 9) shows that the spectral efficiency for WiMAX is expected to be about three times that of basic HSDPA in both the uplink and the downlink. From other published sources (Ref. 12) this seems to be close to the consensus view (with some analysts suggesting a factor of only twice the advantage in the downlink). Again these figures are for basic WiMAX and R5 HSDPA. WiMAX will benefit from the use of smart antennas which will raise the efficiency by a further factor of two to four

	DL Throughput per channel/ sector Mbit/s	ULThroughput per channel/ sector Mbit/s	DL Spectral efficiency Bit/s/Hz	UL Spectral efficiency Bit/s/Hz
1XEVDO Rev. 2*1.25MHZ	1.2	0.4	0.8	0.4
3XEVDO Rev.B	4.4	1.6	0.95	0.25
HSDPA	4	0.6	0.7	0.15
HSPA	4	1.7	0.8	0.35
802.16e SIMO symmetric frame split	6	3.2	1.1	0.7
802.16e MIMO symmetric frame split	9	4	1.6	0.9
802.16e SIMO 3:1 Asymmetric frame split	9	1.8	1.2	0.6
802.16e MIMO 3:1 Asymmetric frame split	14	2	1.8	0.8

*Table 5-5. Mobile WiMAX is 1*10MHz, HSDPA, HSPA and EVDO Rev B are 2*5MHz. and 1xEVDO Rev.A is 2*1.25MHz. All figures from WiMAX Forum modelling. 3:1 frame split means the downlink sub-frame is three times longer than the uplink sub-frame.*

in dense urban environments and HSDPA will be enhanced with better receivers and diversity transmitters to gain a factor of 2. HSUPA will also improve uplink efficiency – though not by anything like as much as HSDPA has for the downlink.

In terms of throughput for individual users WiMAX can reach data rates of 75 Mbit/s using 64 QAM and ¾ coding in a 20MHz band (Ref. 13). This is what the salesmen will quote! But it is not representative and will apply only to mobiles close to the base-station where interference is mitigated, there are no other users active and a large following wind. However, typical quoted maximum user rates for WiMAX are about 9 Mbit/s downlink and 1.3 Mbit/s uplink (for a 3:1 downlink: uplink split in a 10MHz TDD band). Ref. 13 (based on WiMAX forum) gives a WiMAX sector throughput of 14 Mbit/s down and 2 Mbit/s up – which are very similar to the figures I have given above. Ref. 12 gives a peak WiMAX rate of 23 Mbit/s up and 4 Mbit/s down (10MHz 3:1 downlink: uplink split) – which are almost certainly PHY layer rates – real applications would see rates 20–40% lower than this level.

When you add in smart antennas (phase 2 WiMAX – 2010/11) these rates will increase to between 12Mbit/s down and 1.8 Mbit/s up (macro cell for a 3 : 1 downlink: uplink split in a 10MHz TDD band) and 16Mbit/s and 2.2Mbit/s for a micro cell (where MIMO is favoured due to the rich multipath environment). Ref. 12 gives the improvement form phase 1 to phase 2 data rates as a two times improvement for downlinks.

Sprint (in Sept. 2007) demonstrated a modem/notebook working with WiMAX on the Chicago river offering 3.2Mbps downstream and 1.5Mbps upstream with a 70ms latency when in the dock (reducing to 2.4 Mbps down and 1.4 Mbps with 99 ms latency when moving – presumably slowly!)

The next most important performance metric of a wireless network is the latency. Typical figures quoted are around 70–100 ms for wave 1 products and 50 ms or so (one way) in the second phase.

Set-up time – meaning the delay from starting an application to getting IP packets flowing is also much reduced in WiMAX – with typical figures of 100–150 ms being suggested. This compares with 1–2 sec for R5 HSDPA and 650 ms for HSUPA R6. However, there is an option in UMTS to keep the PDP context permanently active and this also reduces set-up times to 300 ms or so. Where this would be of benefit is for applications that send packets sporadically but where latency is also an issue – such as in some multiplayer games.

So, in summary, there is some evidence (a mixture of measurements and modelling) that suggests WiMAX will be a significant (i.e. at least twice as good in speed, capacity and efficiency) improvement over HSPA. MIMO will probably double this advantage again. Of course these improvements will be gradual and will, like HSPA, be introduced in phases.

5.11 Conclusion

WiMAX is a fully-fledged cellular mobile system – with all the bells and whistles of: handover, QoS, security, idle-mode support and so on. However, it is not the product of the usual industry standards bodies but rather it has been developed by the computer/Internet industry. It has two technological "magic bullets" – OFDM and multiple/smart antenna technology although these are being integrated into 3GPP and 3GPP2 standards with HSPA+ and LTE. WiMAX is on a different timescale and may well appear before LTE – giving it a window of opportunity.

Undoubtedly WiMAX will be used in different regions for different purposes – in some areas it might serve as an alternative to fixed lines for Mobile Broadband and in others it might offer a wider range of services such as those offered by WiBro in South Korea today:

– VoIP;

– Video calling;

– IMS services;

- Multimedia Conferencing;

- Presence;

- PTT (Push to talk, also push to video);

- E-commerce;

- Location based services (e.g. traffic info);

- Instant Messaging;

- Networked Games;

- IP-TV;

- "Information on Demand" (IoD);

- Web access.

Will WiMAX be successful? It might be in some markets. It seems likely mobile WiMAX will be deployed and marketed in different countries for different things. In some places it might succeed as a super-WiFi in others it might become a DSL replacement. It will also succeed where existing mobile operators (such as KT and Sprint-Nextel) take it up and deploy it alongside existing technologies.

In the end it comes down to three things – spectrum, terminals and applications. If the spectrum is available at the right frequency, in the right amounts and at the right price and if terminals appear rapidly, with low cost chip sets, then WiMAX has a good chance. If applications and services that need lower cost, higher bandwidths and high volumes become popular then WiMAX will have a sufficient advantage over 3G to encourage take up. Existing DSL and 3.5G deployments will raise the barriers to success and markets with low DSL penetration and less developed 3.5G networks will, conversely, accelerate success.

References

[1]   *Business Week* Online: 26 Aug. 2005.
[2]   http://www2.sprint.com/mr/news_dtl.do?id=12960
[3]   Geoff Haigh BT – talking about WiMAX.
[4]   WiMAX, L. Nuaymi, John Wiley & Sons Ltd, 2007, 879-0-0470-02808.
[5]   "OFDM for Mobile Data Communications", Whitepaper, March 2003, Flarion Technologies, Inc. www.flarion.com
[6]   http://news.zdnet.co.uk/communications/0,1000000085,39287701,00.htm
[7]   "MIMO systems and transmit diversity", http://www.comm.utoronto.ca/~rsadve/Notes/DiversityTransmit.pdf.
[8]   "Executive Summary: Mobile WiMAX Performance and Comparative Summary" 118KB .PDF - Sept. 2006 from http://www.wimaxforum.org/technology/downloads/Mobile_WiMAX_Performance_and_Comparative_Summary.pdf.

[9]   WiMAX forum www.wimaxforum.org/
[10] *Business Briefing*: Wireless Technology 2004.
[11] *Business Week* Online: 26 Aug. 2005.
[12] "EDGE, HSDPA and LTE", RySavy Research, Peter Rysay, Sept. 2006.
[13] WiMAX – Chris Beardsmore, lecture as part of "Wireless Communications and Future Applications WLAN, Bluetooth, Ultra-wideband, WiMax and 4G", University of Oxford, June 2007.
[14] "Sprint Nextel switches on WiMAX lights", 17/12/2007 http://www.wimax-vision.com/newt/l/wimaxvision/article_view.html?artid=20017489334.

More to Explore

"WiMAX", L. Nuaymi, Wiley, 2007, 879-0-0470-02808 – very technical and contains all the detail on the standard – weak on commercials or real deployments.

"Mobile WiMAX-Part I: A Technical Overview and Performance Evaluation", WiMAX Forum, August 2006, www.wimaxforum.org. (NB: there are many other useful whitepapers here)

"Mobile WiMAX-Part II: A Comparative Analysis", WiMAX Forum, August 2006, www.wimaxforum.org

Hassan Yaghoobi, "Scalable OFDMA Physical Layer in IEEE 802.16 WirelessMAN", *Intel Technology Journal*, Vol. 8, Issue 3, August 2004, pp. 201–212.

www.WiMAX.com – excellent site on deployments and news.

Chapter 6: Convergence and the IMS

6.1 Introduction

Having looked at a great number of new wireless technologies – from HSPA to WLAN to WiMAX to LTE, it is a relief (both for the reader and writer!) to talk about something else. Convergence is a much overworked term – some people say it is when you get some permutation of: mobile phone, broadband, fixed phone and pay-TV from a single supplier (the so called quad play). Some people say it is when you get one bill for all these things. To date, however, that has pretty much been the limit of convergence with fixed and mobile world solutions existing as stovepipes. Take email – until recently if you wanted to do that on a mobile you had to buy a special terminal and subscribe to a special service (Blackberry). Instant messaging is the norm on the fixed Internet but has been very slow to appear on the "Mobile Internet" – partly because SMS is so lucrative. How do you manage your calls? If somebody wants to contact me how do they do it? Well I suppose if it is a work call they might ring my work mobile – no answer? – ring my work fixed phone, no answer? – send an email. . . and so forth and that is not taking into account that the whole shooting match is reproduced for my private phones and email. The problem is partly technical – a mixture of legacy systems (PSTN, IN etc.) and partly commercial in that convergence will mean winners and losers and nobody wants to be a loser! Imagine if the mobile Internet was as cheap as the fixed Internet – and supported VoIP!

The whole term convergence begs the question – convergence between what? Traditionally this would be fixed and mobile but increasingly it is now

IP for 4G   David Wisely
© 2009 John Wiley & Sons, Ltd.

being extended, in a somewhat meaningless way, to mean convergence of everything – identities, services, terminals – everything. Certainly the days of senior executives carrying: laptop, mobile phone, PDA, Blackberry, camera and MP3 player with them are numbered.[1] In other contexts convergence can mean convergence of everything towards the Internet solution.

In this chapter we will look mainly at the IMS – the Internet Multimedia Subsystem. This is a sort of "Internet in a box" – that was conceived as a way of offering all sorts of whizzy IP services (VoIP, Instant Messaging, video calling...) over the packet-switched part of the 3GPP mobile network. The idea was for mobile operators to offer these – in conjunction with appropriate Quality of Service – to their customers. The authentication and billing already in place for network access would be re-used at the service level. There are interfaces from the IMS for third parties to create service within this controlled environment. The deployment of the IMS by cellular mobile operators has not quite happened – held back by the poor performance of 3G packets services that have limited the scope for real-time services needing QoS and by the continuing growth and low cost of circuit-based mobile voice. However, there are two areas where the IMS is being seen a vital development: Convergence and WiMAX. Operators with both fixed and wireless networks are looking at the IMS to create an access-independent platform for service creation over and above the "bit-pipe" level. The IMS is also important as it is the basis of ETSI's Next Generation Network (NGN) that extends the IMS concepts to apply to DSL and WLAN.

Having gone through how the IMS works and what it might do for you we will then look at real convergence today – some of the innovative products that have emerged in the last couple of years – such as BT's Fusion phone that connects to both GSM and WLAN and seamlessly switches between them. Then we will look at alternatives to the IMS.

6.2 The IMS

So what exactly is the Internet Multimedia Subsystem? And why was it standardised? One way to answer these questions is to go back to where Chapter 2 left off and look at the functionality and services that current (R99) UMTS networks offer. Figure 6.1 shows an outline R99 network – with the exception of the RAN this is also nearly identical to a GPRS network. As was explained in Chapter 2 this was the "revolutionary air-interface, evolutionary network" approach that 3GPP decided upon for the introduction of 3G.

---

[1] I can't help recalling my old boss Peter Cochrane – a legend in his time as head of Research at BT – not least because he once sent me an email saying "you guys make me vomit" – I still have it!. Peter was a gadget and Internet pioneer and carried a toolkit to dismantle foreign phone sockets to enable internet access anywhere he was travelling. He used to carry a whole suitcase of gadgets but then put his shoulder out with the weight!

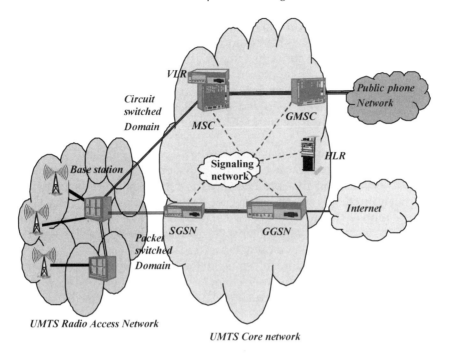

Figure 6-1.   Typical R99 UMTS network. (*Source*: BTTJ. Reproduced by permission of © British Telecommunications plc.)

When the R99 network was being standardised – in the late 1990s, obviously – the Internet was really taking off in a big way for consumers and businesses: people were experimenting with VoIP, buying books, accessing online auctions and so forth. It was apparent to mobile operators and manufacturers that the Internet business model was very different from the mobile value chain. If I buy a book from Amazon on-line several players are involved: My phone line provider (BT), my ISP (AOL), the vendor (Amazon) and the payment authoriser (Visa). In contrast when I buy a mobile ring-tone it tends to be from the Vodafone Live portal and there is only one player – Vodafone. In addition the Internet was offering services that went beyond the simple client-server model of mobile internet offerings – there were store and forward (e.g. email) and peer-peer (e.g. Instant Messaging) services beginning to become popular on the net.

3GPP members realised that R99/GPRS networks would only ever offer a bit-pipe connection to the Internet and that the data services that they were offering, WAP and SMS, would always be a very poor relation to the Internet experience on a fixed PC. They feared that users would simply buy connectivity from them the way they do from fixed ISPs and that all the value in the value chain would "escape" to the likes of: Amazon, Visa, Microsoft and Google. Evidence for this is clearly provided by the way operators around Europe have either prevented or made difficult "off-portal" browsing – i.e. visiting mobile websites not approved of (i.e. paying a fee to) or not provided by the mobile operator. T-mobile have only recently (late 2005 – Ref. 1) introduced "Web and Walk" and 3 at about the same time have just opened their "walled garden" to non 3 sites on an "approved" list. Further illustration of this point is how hard it has been to get email on a mobile – you either had to shell out for a specialist, stovepipe device and service (a Blackberry) or pay usurious, per Mbyte, download rates for a PCMCIA data card that allowed open Interne/Intranet access.

In addition there was a great interest in VoIP – with early SIP services showing that it was possible to offer greatly enhanced IN services for multi-media calls/ sessions and that presence was a very popular service. Operators were interested in whether a packet network could support voice – the biggest problem being reducing the latency from 800ms (GPRS) and 200ms (R99) to nearer 50ms – and whether this would offer cost saving and new service opportunities (e.g. presence-based services).

From all this thinking emerged the IMS: a complete sub-system that could be added, with minimal changes, to an R99 network to offer a complete multimedia services creation platform. Among the design goals of the IMS were:

- To support real-time services (e.g. voice and video) over a packet-switched network – which might be more efficient and economical, when aggregated with other IP data, compared to circuit-switched options.

- To support the business case for the introduction of QoS and the means to charge for services requiring QoS support.

- To offer presence services (e.g. Instant Messaging).

- To support multiple IP and Telephone identities for each user.

- To allow interfaces for third parties to create value-added services (e.g. the smart routing of calls based on a person's calendar).

- To allow roaming users to get access to the same services they would when at home – with proper billing and roaming arrangements.

- To extend the cellular strong security solution (i.e. SIM cards) to authenticate users to service providers and support high-value services and purchases.

- To allow operators to charge for specific data services – as opposed to simply by volume (e.g. per packet or MByte) or by time.

Some of these reasons are commercial, some are technical and related to cost-saving and others relate to new service creation – the rationale for the IMS is a mixture of all of these and we will return to evaluate how successful the IMS is at achieving these goals in the final section of this chapter.

Originally the IMS was to be included in a version of the 3GPP standards to be known as R2000 (Release 2000). Unfortunately it was quickly realised that it

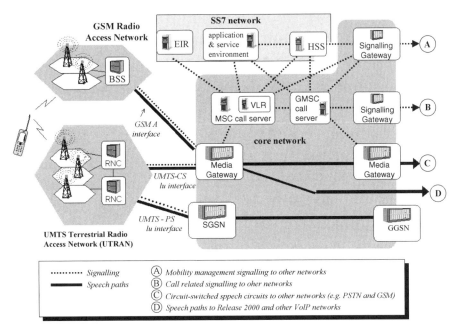

Figure 6-2. 3GPP Release 4 – offering VoIP in the core network. (Source: BTT*.
Reproduced by permission of © British Telecommunications plc.)*

would take much longer to complete work on an all-IP UMTS and the standard was split into two parts – R4 and R5. R4 offers operators support for VoIP in the core network – allowing voice to be transported over IP whilst being controlled by a call server (Figure 6.2). The call server (soft switch) is the old (3G) MSC with the switching function moved remotely to a Media Gateway (MGW), this being controlled by the call server using MEGACO(Media Gateway Control Protocol). The advantages of this might include separately dimensioning the control and transport elements of the MSC and savings on the core network transport costs by aggregation of all traffic, voice and data, as IP.

Release 5, which completed standardisation in 2004, adds the IMS and full support for all services to be operated over the packet network. Well, almost! As we will see later, R5 has no solution for emergency calls and so R5 terminals will all include circuit-based voice for emergencies and roaming to R99 networks. Since then we have moved on to Release 8 which is just being finalised in 2008 – each subsequent release has added to the IMS but the basic operation is unchanged.

Figure 6.3 shows an outline of how the IMS fits into the R5 architecture – it is basically an overlay (i.e. a subsystem). Users will get a new terminal that can "talk" to both the IMS and the R99 network. If VoIP and real-time services are being offered the SSGN and GGSN have to be upgraded to allow QoS authorisation and low-latency support for IMS services (and, by implication,

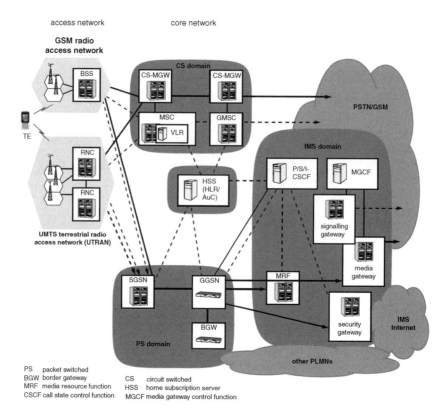

access network   core network

Figure 6-3. How the IMS forms part of an R5 UMTS network (courtesy BTTJ). (Source: BTTJ. Reproduced by permission of © British Telecommunications plc.)

not for non IMS services!). In fact the latency on most current live 3G networks – 200–300 ms (Ref. 2) – does not support real-time services and upgrades will be needed to allow real-time voice operation. With HSDPA and HSUPA this is starting to happen as was described in Chapter 4 and the author's lab have had success running VoIP clients over HSPA links with fairly good results. Operation of the IMS is based on SIP – with a user agent on the terminal and the core of the IMS basically acting as a SIP proxy server. This is similar to most current presence and VoIP services on the Internet. Essentially the IMS performs like a carrier-grade SIP proxy server with added features:

- Strong security – based on SIM cards – with an enhanced HLR now called the Home Subscription Server (HSS).

- A break out for VoIP to non IP networks (e.g. the PSTN) – involving signalling (SGW) and media gateways (MGW).

- Interfaces for operator and third party application servers to offer IN-like functions to call/session initiation and routing.

- Special functionality for multi-party calls/session (essentially an IP bridge) –
the Media Resource Function (MRF).

In this section we will start with a complete description of how an IMS user makes a VoIP call to another IMS user. Section 2.1 will cover the important issues of: User Identity, Registration, SIP message flows and QoS-authorisation. Section 2.2 then deals with how the IMS authorises QoS for IMS-originated services; Section 2.3 deals with presence and the creation of services based on the IMS; Section 2.4 looks at roaming – both to visited IMS and non IMS networks. A commercial summary of the IMS and possible alternatives is held back for the conclusion – after we have looked at non IMS convergence technologies.

There are a number of key aspects of the IMS which represent major decisions that have ramifications on the architecture and operations of the IMS. In particular:

- The decision to use SIP as a call/session control protocol;

- The re-use and extension of IETF protocols;

- The decision to mandate the use of IPv6 in the IMS;

- The fact that all services are controlled by the home network IMS – even when roaming.

We will discuss the reasons for these decisions and the implications – both positive and negative – as they arise throughout this section. The only prerequisite to read about how an IMS works is a knowledge of SIP – Session Initiation Protocol – the well-known IETF protocol used for VoIP, video calling, IM and many other things in the Internet. If you don't know SIP don't worry – you can download my handy tutorial – an article and slide set (www.wisely.org/consult/SIP) for free.

6.2.1 User Registration

6.2.1.1 Attachment to the IMS

In this section we follow an IMS user, equipped with a new IMS handset, making a VoIP call to another IMS user on a Release 5 network – the very simplest scenario possible for IMS operation.

Firstly, our user must switch on their terminal and connect to the UMTS network in exactly the same way that an R99 terminal connects and authenticates, as described in Chapter 2. A USIM application on the user's smart card (UICC) mutually authenticates to the HSS in the normal way and sets up encryption keys for circuit voice use and the user is free to make circuit-based voice calls at this point (an IMS terminal must also be capable of circuit voice for roaming and emergency calls). In our example, however, the user wishes to make a VoIP call and so must register with the IMS first.

One of the original ideas – and probably, with hindsight a poor choice – was that the IMS must use IPv6 and that IMS-capable terminals would be either v6 only, if the underlying transport network had been upgraded to v6, or would be dual (v4, v6) stack to allow packets to be sent on a v4 packet network whilst communicating with the IMS using v6! The idea was that by the time an IMS would be deployed (about 2006) most networks would have converted to IPv6. It actually turns out that IPv6 is a bit like nuclear fusion – always five years away! Since IPv6 shows no signs of taking off any time soon the leading IMS vendors are producing IPv4 versions of the IMS and so avoid all the issues of interworking with legacy IPv4 networks – such as the Internet! Originally an application layer gateway was included to perform the very problematic IPv4-IPv6 translation for the signalling traffic – mercifully this clunky solution looks like being unnecessary.

In order to register with the IMS the terminal needs to set up a PDP context which will carry the SIP signalling from the terminal to the IMS via a GPRS attach procedure as previously described in Chapter 2. An *Activate PDP Context Request* message is sent from the terminal to the SGSN requesting connection to the IMS (by means of an appropriate APN – Access Point Name – that points to a Release 5 capable GGSN – which would be separate from the R99 GGSN). The R5 GGSN is responsible for allocating the terminal an IP address in the *Activate PDP Context Accept* Message. Note that in the standards the terminal is sent a 64-bit prefix with which it is free to choose a 64-bit suffix to create an IPv6 address. The PDP context is always of the conversational class and so gives the IMS signalling traffic priority over traffic in the other three UMTS QoS classes.

The IMS terminal consists, in its simplest form, of:

- A standard R99 radio interface and stack;

- A SIP user agent;

- An ISIM application on the UICC;

- New Codecs for voice and video over IP;

- VoIP and IP video applications.

3GPP debated for a long time about the call/session control protocol that would be used to initiate, modify and terminate IP sessions. The debate boiled down to a choice between the IETF SIP protocol and H323, a heavyweight VoIP protocol, from the ITU (i.e. from the telecoms world). H323 covered all aspects of setting up a VoIP network – including media transport and QoS – jobs that were done by several IETF protocols and was something of a stovepipe solution. The choice of SIP – which at the time was just emerging as an important peer-peer protocol was the first time a protocol not defined in the telecoms standardisation bodies was specified for a telecommunications service.

However, the basic SIP RFC (3261) has been greatly extended by negotiation between the IETF and 3GPP to make it suitable for use in the 3GPP IMS and the SIP user agent on the terminal is somewhat different from the typical SIP User agent (as, for example, found in MSN Messenger). In fact several other IETF protocols were incorporated into the IMS and these will be explained later, as they arise.

Having set up the PDP context towards the IMS the terminal needs to discover the IP address of a SIP proxy server in the network. This procedure has been incorporated into the Activate PDP Context Procedure – so that the SIP user agent on the IMS now has the IP address of a SIP proxy server. We shall see, later on, when we consider other access networks, such as WLAN and WiMAX, that the IMS can attach to such networks (called ICANs – IP Connectivity Access Networks) with some modification. In such cases the IP address of the SIP server is supplied by a modified DHCP procedure.

Now things get complicated! As there are three "flavours" of SIP proxy server (called Call Server Control Function – CSCF as shown in Figure 6.4) in the IMS network:

- P-CSCF – Proxy Call Server Control Function;
- I-CSCF – Interrogating Call Server Control Function;
- S-CSCF – Serving Call Server Control Function.

The first CSCF that the terminal interacts with is the Proxy (P-CSCF). This performs a number of functions that we will cover in more detail as the call set-up is explained:

- SIP message compression between the P-CSCF and terminal – mostly to speed up the SIP signalling rather than to save bandwidth.

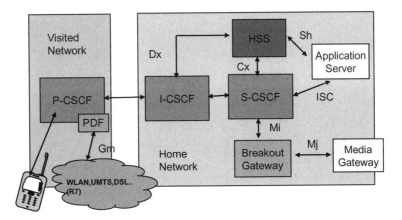

Figure 6-4. IMS major elements and interfaces. (*Source*: Author. Reproduced by permission of © Dave Wisely, British Telecommunications plc.)

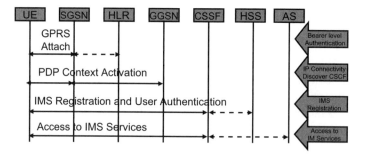

Figure 6-5. Overview of the IMS registration process. (*Source*: Author. Reproduced by permission of © Dave Wisely, British Telecommunications plc.)

- Encryption and message integrity between the P-CSCF and the terminal.

- Validation and verification of SIP messages sent by the terminal.

- Charging records of sessions to and from the terminal.

- Validation and verification of the user identity (to the rest of the IMS).

Having discovered the IP address of the P-CSCF the terminal attempts to register with the IMS, only then is access to IMS services granted as shown in Figure 6.5.

6.2.1.2 IMS-level Registration

IMS-level authentication is basically the procedure by which the network authenticates the user and then authorises them to access the IMS services. The user is authenticated by means of a SIP REGISTER message sent to the P-CSCF – this is similar to a standard SIP Registration but modified by 3GPP and containing many extra pieces of information in an attempt to speed up the whole process and save round trip SIP messages.

In order to understand the registration procedure it is necessary to know how the IMS identifies users by means of public and private identities. A user has a number of public identities such as:

- SIP: dave.wisely@bt.com;

- SIP:++44-1473-643784@bt.com;user=phone.

Both identities are in the form of SIP URI (the IMS does not allow TEL URI – i.e. bare telephone numbers – to be used). A user can have a number of such identities – for example for private and business use. A user also has a private identity – which takes the form of a NAI (Network Address Identifier – e.g. 17356trf7dfgutf@bt.com). The public identities are directly analogous to

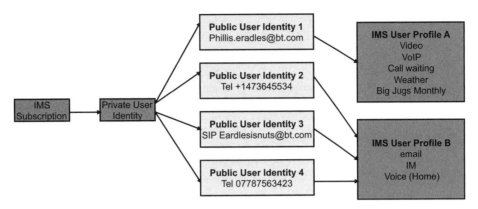

Figure 6-6. Public and private identifiers in the IMS. (*Source*: Author. Reproduced by permission of © Dave Wisely, British Telecommunications plc.)

MISDNs (i.e. mobile phone numbers) and the private identifier is analogous to the IMSI. Figure 6.6 shows the relationship between these identifiers.

The SIP user agent on the terminal reads the private user identifier from the I-SIM application running on the UICC and selects (e.g. the user selects) the public identity (or identities) that are being registered and, together with the terminal IP address and the home domain identifier (e.g. bt.com – read from the ISIM) are composed into a SIP register message and this is sent to the P-CSCF as shown in Figure 6.7. In general the P-CSCF may be located in the visited network and will use DNS to resolve the home domain identifier to the IP address of another type of CSCF – the Interrogating or I-CSCF. The I-CSCF sits at the gateway to the home network and acts as a firewall to hide the

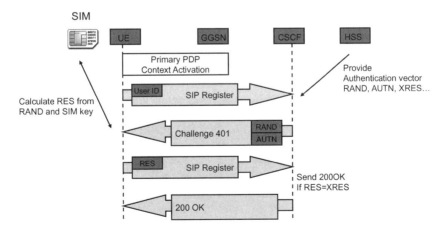

Figure 6-7. IMS registration procedure. (*Source*: BT. Reproduced by permission of © British Telecommunications plc.)

network details. The I-CSCF does not keep state – since there are typically more than one and the DNS load balances across them – and so needs to contact the HSS to see if the user is already registered. Using the IETF AAA protocol DIAMETER the I-CSCF discovers that the user is not registered and retrieves the user profile. At this point the I-CSCF must allocate the user a Serving CSCF (S-CSCF) to actually provide any service. Confused? – I am not surprised – but the S-CSCF is the actual SIP proxy server that will provide service to the user – e.g. holding a voice call state model and triggering presence services, for example. The I-CSCF is just a stateless firewall at the gateway to the network that hides the network and allows different users to be allocated different S-CFCFs based on the user profile and the capabilities of the different S-CSCFs – for example some will support presence, some push to talk and some will be voice only.

The I-CSCF makes a decision as to which S-CSCF the user will be allocated to and proxies the SIP REGISTER message. The S-CSCF contacts the HSS using DIAMETER to obtain authentication vectors (the usual SIM-like challenge-response we saw in Chapter 2) and to record the S-CSCF address in the HSS so that incoming and future outgoing session signalling passes through the allocated S-CSCF (which does not change until the user de-registers). The S-CSCF sends back a SIP 401 (Unauthorised) response and includes a challenge based on the mutually secret key hidden in the HSS and UICC, identified by the private user identifier. Figure 6.7 shows the challenge response – which is completely analogous to the USIM challenge described in Chapter 2, although the algorithms and keys are completely separate. The SIP user agent sends the challenge to the ISIM, receives a response and sends this back to the P-CSCF in another SIP REGISTER message. This is forwarded to an I-CSCF (not necessarily the same one as before) and again the I-CSCF contacts the HSS – however this time the HSS has the address of the allocated S-CSCF and the I-CSCF forwards the message on. If the S-CSCF confirms that the challenge response is correct it contacts the HSS to complete user registration and download the user profile. Figure 6.8 shows the basic information flow and Figure 6.9 the situation immediately after registration. Yes, it is a lot of information but look at Figure 6.9 and you will see that it just means the terminal has been allocated a serving SIP proxy in the home network. The HSS knows which S-CSCF and how to contact the terminal and that the terminal and proxy SIP server have a secure relationship!

One of the by-products of the registration is that the P-CSCF and SIP user agents derive cipher and integrity keys which are used to set up an IPSEC tunnel from the terminal to the P-CSCF. No further encryption is normally used as the internal networks are considered secure.

One further point worth noting about registration is the problem of knowing when a register server shuts down (e.g. because it fails). In standard SIP there is no way for a SIP UA to know that it is not reachable. 3GPP have extended SIP (in RFC 3680) to allow an IMS terminal to SUBSCRIBE to its own registration state and be alerted when it was not reachable.

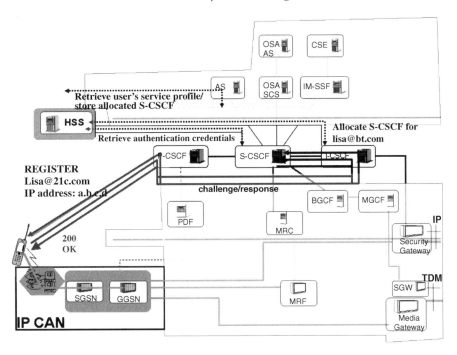

Figure 6-8. basic SIP message flow in IMS registration. (Source: BT. Reproduced by permission of © British Telecommunications plc.)

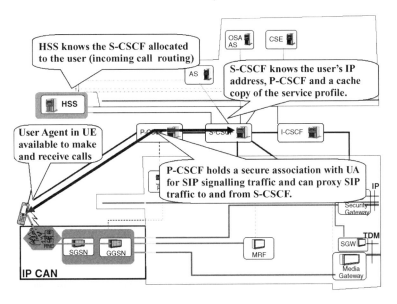

Figure 6-9. Information distribution after IMS registration is complete (the CAN is the Connectivity Access Network – UMTS, WLAN etc.). (Source: BT. Reproduced by permission of © British Telecommunications plc.)

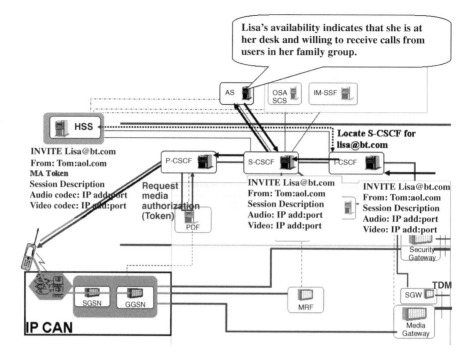

Figure 6-10. *SIP INVITE received by bt.com.* (*Source*: Maria Qauvas. Reproduced by permission of © British Telecommunications plc.)

6.2.2 Basic IMS Session Set Up

6.2.2.1 The INVITE Message

Figure 6.10 shows the basic SIP message flow when Tom (Tom@aol.com – one of Tom's public identifiers) tries to contact Lisa (Lisa@bt.com) for a VoIP session. Assuming Tom is on his home network and registered and that Lisa is on her home network (bt.com) and registered then Tom can make a VoIP call attempt. His SIP user agent creates a SIP INVITE message and sends it to his P-CSCF – usually referred to as the originating P-CSCF. In fact the P-CSCF and S-CSCF stay in the loop of all SIP messages sent by means of SIP extension (the Record-Route header) and loose source routing. In basic SIP the proxies can drop out after the initial INVITEs are delivered and the end parties discover their respective IP addresses but in the IMS this is not allowed (or it would be hard to control service creation and charging and the P-CSCF provides compression and encryption and the S-CSCF is the trigger point for any extra services). Both originating and terminating P-CSCFs and S-CSCFs stay in the signalling loop but the terminating I-CSCF sees only the INVITE and all its responses, but not other requests such as PRACK, ACK and BYE – dropping out of the chain.

Tom's INVITE is read by his S-CSCF and this then uses DNS to determine the IP address of an I-CSCF in the bt.com domain and forwards the INVITE. The bt.com I-CSCF looks up Lisa@bt.com in the HSS to determine if she is registered and, if so, which terminating S-CSCF has been allocated to her. In Figure 6.10 the trigger feature is shown – whereby SIP INVITE messages can be used as triggers for services – in this case to screen calls (Is caller on family list? If so pass INVITE back to S-CSCF, if not then, for example, send back a SIP 302 – moved temporarily – response pointing at a voice mail). We will look at such services in much greater detail later in the chapter. Lisa's S-CSCF forwards the invite on to her but this does not yet make her phone ring! It may be that there are no resources to support a VoIP call in either hers or Tom's radio networks – so first QoS resources must be reserved for the call.

The INVITE message would be something like that shown in Table 6.1. This INVITE is fairly easy to read – the INVITE is to Lisa and comes from Tom's user

---

INVITE sip:lisa@bt.com SIP/2.0
Via SI/2.0/UDP [132.146.34.1]:5059 comp=sigcomp; branch = gdududdw
Max-Forwards: 70
Route <sip:pcscf1.aol.com:5058;lr;comp=sigcomp>,<sip:scscf3.aol.com:5058;>,
P-Preferred-Identity: "Tom Cobly" <sip:toma@aol.com>
Privacy: none
P-Access-Network-Info: 3GPP-UTRAN-FDD; utran-cell-id-3gpp= HGFYSV46I
From: <sip:tom@aol.com>
To: <sip:lisalargejugs>
Call ID: TY56
Cseq: 45 INVITE
Require: Precondition, sec-agree
Supported 100rel
Security-Verify: ipsec-3gpp;q=0.1; alg=hmac-sha-1-96;
spi-c=7847595;spi-s=7264247;port-c=5057;port-s= 5058
Contact:<sip: [132.146.34.1]:5059 comp=sigcomp>
Allow: INVITE, ACK, CANCEL, BYE, PRACK, UPDATE, REFER, MESSAGE
Content-type: application/sdp
Content length:356

(SDP PART)
v = 0
o = 132.146.34.1
s = -
c = 132.146.34.1
t = 0
m = audio 8283 RTP/AVP 97
b = AS:25.4
a = curr: qos local none
a = curr: qos remote none
a = des: qos mandatory local sendrecv
a = des: qos none remote sendrec
a = rtpmap:97 AMR

Table 6-1. SIP INVITE from Tom to Lisa as sent from Tom's user agent.

agent recorded in the Via field. The route field already contains the originating (i.e. AOL) P-CSCF and S-CSCF which Tom learnt during PDP attach and IMS registration procedures respectively. The preferred identity will be validated by the P-CSCF as a valid public id associated with Tom's private id. The P-CSCF will change the field to P-Asserted identity to show this to the rest of the network. Some access-network specific radio id is allowed – but note, this is stripped out before the INVITE leaves the home network (it might be used to provide location-based services, say).

The from, to and call-id fields are end-to-end fields and not read by the CSCFs or verified by them (this is actually a feature of standard SIP). Two SIP extensions are mandatory: pre-conditions and security agreement (noted in the "require" field). We will explain pre-conditions shortly as this is related to QoS set-up. Security agreement is essentially the IPsec association between the P-CSCF and the terminal and the details are given in the Security-Verify field. The contact header is the IP address and port number of Tom's user agent – showing the SIP compression used between the UA and the P-CSCF. The allow field shows the (extended) methods that the UA supports and we will see some of these shortly. Finally, there is a SDP description – the offer from Tom – of the desired session – note that 3GPP INVITE messages only carry SDP bodies i.e. not other protocols that SIP is capable of carrying. The example in Table 6.1 shows a request for an audio session on port 8283 using the (mandatory in 3GPP) AMR codec (no. 97).

The message is sent by Tom's UA to the originating (AOL) P-CSCF which:

- Checks that the INVITE has come from Tom's terminal – using the security association set up at registration.

- Checks the INVITE.

- Validates the SDP offer according to the general network policy– some codecs (e.g. G711 (64 kbit/s)) may not be allowed on the network because they take too much bandwidth.

- Validates the p-preferred id and replaces it with a p-asserted id.

- Records itself in a Record-Route field – so that it is always visited.

- Begins charging function.

The originating P-CSCF proxies the INVITE to the terminating S-CSCF (using the routing information in the INVITE) – normally there is no need for further security between these CSCFs. The S-CSCF downloaded the user profile during registration and this includes a set of initial filter criteria – we will explain in detail later – but basically describes conditions under which the INVITE will be sent to a further SIP server for service creation. For example Tom may subscribe to a fancy service that allows Lisa to see a picture of him on her phone when he is calling. In this case we assume none of the filter criteria match – or Tom is too tight to pay for them. The S-CSCF further polices the SDP request with any user-specific policies – i.e. G711 may be an allowed codec but users might be expected to pay for the extra bandwidth and quality. The S-CSCF then uses a number of DNS queries to determine the IP address and transport protocol support (UDP/TCP) of an I-SCSF in the terminating network (bt.com) – these are cached to improve speed of set up. The terminating I-CSCF needs to locate the S-CSCF allocated to Lisa – so the I-CSCF queries the HSS with a Diameter Location-Information Request (LIR) – the HSS is able to search using public ids as a key and either responds that the user is not registered (i.e. the phone is off or out of range) or proxies the INVITE to the S-CSCF. The I-CSCF may or may not stay in the loop for signalling – depending on whether the operator is happy to expose the S-CSCF to the originating network.

The terminating S-CSCF receives the INVITE and checks the request against the initial filter criteria – in this case Lisa is screening all calls (possibly following some heavy breathing calls from Hairy Bob). The INVITE is proxied to a secondary SIP server (called an Application Server – AS) that either sends the INVITE back if the caller is on Lisa's allowed list or sends a SIP response (such as 302 temporarily moved with the address of an announcement or voice mail service) if the caller is not on the list. The INVITE is proxied to the terminating P-CSCF. The P-CSCF checks if the privacy field is set – if so it removes the caller's id from the INVITE so that the callee is not able to see the caller's identity. In addition the P-CSCF adds a token to allow the terminal to request the appropriate network QoS for the media – as we will see in the next section.

Finally, Lisa's terminal receives the INVITE. But we are still a long way from ringing the phone. The pre-conditions extensions means that the terminal must respond to the SDP offer with a SIP 183 Session Progress response (see Figure 6.11).

6.2.3 IMS QoS Authorisation

The 183 Session Progress provisional response is part of the pre-conditions SIP extension and will be acknowledged with a PRACK (i.e. provisional ACK) SIP message. Without reliable acknowledgment of provisional responses (the standard SIP approach) it is possible the response will be lost when transmitted over a UDP segment. The 183 response details those parts of the SDP offer that are supported at the callee's terminal (e.g. the voice and video codecs), in this case Lisa's terminal supports the AMR voice codec as it is mandatory but Tom might have added G711 in the hope of having a high quality voice conversation. The 183 response makes its way (via the terminating P-CSCF (bt.com) and S-CSCF (bt.com) and originating I-CSCF (Aol.com – not shown in Figure 6.11 for clarity!), S-CSCF (Aol.com) and P-CSCF (Aol.com) to Tom's terminal. The SIP UA can now decide on the codecs that will be used and begins to reserve resources for these (in this case just AMR voice) within this leg of the network. Figure 6.11 shows the basic message flows in the call set up.

In order to ensure resources are likely to be available for IMS services the R5 GGSN has been modified to allow priority to be given to IMS services by means of a token that is generated by a Policy Decision Function (PDF – part of the

S-CSCF in R5 but separate in R6) and given to the parties involved in authorised IMS sessions. In this case tokens are sent with the INVITE to Lisa and with the 183 response to Tom and are used with a PDP Context Activation Procedure to establish a PDP context with the R5 GGSN that offers the appropriate QoS for the requested session. Note that this is a secondary PDP context – in addition to the one set up for SIP signalling to the IMS. In this case QoS suitable for VoIP – low latency guaranteed bandwidth – would be selected. The PDP context activation request is forwarded by the SGSN to the R5 GGSN which then validates the token with the PDF and allows (by means of packet filtering) only the authorised packets to leave the network. The protocol used between the GGSN and PDF is the IETF COPS (Common Open Policy Service) protocol (RFC2748).

On receipt of the 183 response Tom's terminal replies with a SIP PRACK (provisional Ack) that includes an updated SDP offer – which will now contain only one (mutually compatible) codec for each media type – in this case AMR for voice and begins to reserve resources in the originating network by requesting a secondary PDP context and including the token it received in the 183 response. The PRACK generates a 200 OK response (Figure 6.11) – note that this relates to the PRACK and not the original INVITE. As soon as Lisa's terminal receives the updated SDP offer it can also start reserving resources in the terminating network by requesting a secondary PDP context with the token received in the Invite message. Once Tom's terminal has completed resource reservation it sends a SIP UPDATE message – which is acknowledged

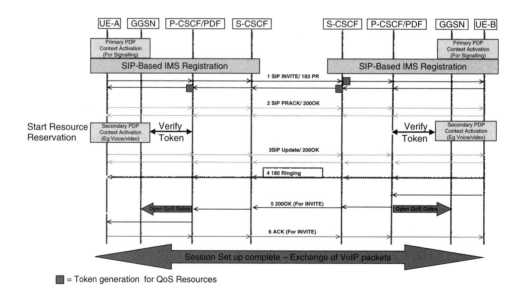

Figure 6-11. Preconditions and QoS authorisation within the basic IMS message flow. (*Source*: BT. Reproduced by permission of © British Telecommunications plc.)

with a 200 OK message. Lisa's terminal now waits for the resource reservation to complete (i.e. the PDP context to be activated) and then, finally, rings the phone.

6.2.4 IMS Service Creation

6.2.4.1 The SIP Application Server (AS)

The IMS would, at first sight, seem quite limited. From what we have described so far an IMS user can make and receive voice, video and multimedia calls to other IMS users in a secure, QoS-enabled environment controlled (and charged for) by the mobile operator. However, that doesn't sound a lot different to current 3G/Skype offerings – so there must be more to the IMS than that. The answer is the use of application servers – essentially different kinds of SIP server and agent that sit in the SIP messaging paths and are able to initiate, modify and terminate a session depending on the service that they are providing. In fact we saw an example during the basic session set up whereby Lisa was able to screen her incoming calls and only receive those from callers on a specific list.

It is early days for the development of additional IMS services but, to give the reader a flavour of what might be available, here are some that have been suggested/implemented/trialled:

Traditional IN services

- Call/Session Screening;
- Short code dialling;
- Call/Session divert;
- Answering service – unified Mail box;
- Click to dial.

Internet Services

- Instant Messaging;
- Presence;
- Push to Talk;
- Divert on busy to web page;
- Personalised tones and images for each caller/callee.

In this section we will start by looking at the rules of how SIP Application Servers are incorporated within the basic messaging framework and how they are triggered. Then we will describe how legacy IN solutions can be incorporated within this framework and, finally, how three of the most important services – Instant Messaging, Presence and Push to Talk – have been standardised to be incorporated within the IMS framework.

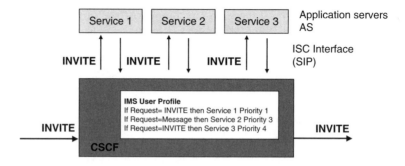

Figure 6-12. The basic SIP Application Server (AS) interfaces. (Source: Author. Reproduced by permission of © Dave Wisely, British Telecommunications plc.)

The basic interfaces for the SIP AS are shown in Figure 6.12 – there is the ISC interface to the S-CSCF which is a basic (i.e. RFC 3261 compliant) SIP interface and the Cx interface to the HSS which uses DIAMTER. Several important points need to be noted here – firstly, that the AS would typically be run by the network operator (BT, Vodafone etc.) and would be in the home network – which always contains the S-CSCF for the user – even when they are roaming. We will see that it is possible for third parties to run a SIP AS, but it seems unlikely that operators will, in practice, allow this (because it exposes their networks without the usual fire walling and load balancing that is always employed when third parties are allowed to attach to networks). The next section shows one example solution that is used in the BT 21C architecture. The Sh interface (see Figure 6.4) allows the AS to download the user's profile as well as "opaque" (i.e. application specific) data that is stored in the HSS but is only understood by the application (e.g. a list of people in my family – which might be used by several ASs and kept centrally). The Sh interface would definitely not be available to any third party AS. Finally, it is important to note that there are two S-CSCFs involved in most sessions – one for the caller and one for the callee and both can offer services. For example, a caller might dial a short code and their S-CSCF might replace it will a full URI and the callee's S-CSCF might offer a voice mail service.

So how are ASs triggered? Each SIP request that passes through a S-CSCF is evaluated against a series of filter criteria and, if a match is found, the SIP message is proxied to the AS. In real terms this means every INVITE, REGISTER, SUBSCRIBE (see later) and OPTIONS message and not PRACK, NOTIFY, UPDATE or BYE message. In fact the most important are REGISTER and INVITE. The filter criteria[2] are downloaded from the HSS as part of the registration phase. A typical filter might look (logically) like this (each line being known as a trigger point):

---

[2] Often called Initial Filter Criteria but, since it was discovered that the proposed subsequent filter criteria broke the rules of standard SIP and were quietly dropped, they are the called only filter criteria.

IF (method = INVITE) AND (Request-URI = sip: weather@bt.com)

OR (method = INVITE) AND (Request-URI = sip: sunnyday@bt.com)

THEN FORWARD TO (server1@bt.com)

In fact the IMS is not limited to just a single filter – there can be a whole list of them – each pointing at different ASs and given a priority number to decide the order in which they are evaluated. This means that a single INVITE might visit several ASs before being finally sent on its way by the S-CSCF. For example a short code might be expanded by the first server and then trigger a second server that recognises the callee as a member of a family group. Of course this can all add delay to the call set up times and the number of potential ASs will be limited in practice. Figure 6.12 shows the case of an INVITE triggering two ASs with different priorities.

So let's follow in more detail the progress of the INVITE to find the weather – the S-SCCF will insert the address of the AS into the Route header and also its own address – to ensure that the INVITE returns via the same S-CSCF. This can be used to store state – e.g. userlisaweather323535r@scscf.bt.com – so that the same AS is not triggered a second time when the INVITE returns! The weather AS then uses the user's location – possibly gleaned from either the P-Access-Network-Info (see Table 6.1) or from the HSS to look up the weather at the user's location (e.g. using the free BBC weather service). In this example the server queries the HSS and discovers that the user is connected on a voice-only terminal and so sends the request to the MRF – the Multimedia Resource Function which can play announcements (e.g. "the weather in Ipswich today will be rain") – the next section gives more details on the MRFC. So the AS proxies the INVITE to ipswich.today.rain@mrfc.bt.com. The AS has finished its job and so does not need to be included in any further requests that belong to the same SIP dialogue and so does not record its address in the Record-Route header (it will still receive any response to the initial request). So an example of a service that needed to stay in the loop might be a translator service – translating English to French in a series of Instant Messages. The S-CSCF detects that the request-URI has changed and doesn't bother to evaluate any further filter criteria and proxies the request to the MRF which replies with a 200OK. The MRF negotiates a codec with the user's terminal and plays the announcement about the rain.

6.2.4.2 Other types of Application Server

In the above example we saw the AS acted as a SIP proxy – in fact it can also act as:

- A SIP User Agent;
- A SIP Re-direct Server;
- A SIP Back-to-Back User Agent.

An example of an AS acting as a SIP UA would be a mailbox that terminated a call by recording a voice message. An AS acting as a redirect server might be a call forwarding or number portability service where the AS responds to an initial INVITE with a 302 Moved Temporarily SIP response. When this is received back by the caller they then make a fresh INVITE with the new URI which might be in a completely different domain.

There are two other types of AS that are included in the IMS standard and these are shown in Figure 6.13. The IM-SSF (IP Multimedia Service Switching Function) is the simplest of these. Towards the S-CSCF it looks just like any other AS – presenting a standard SIP interface – indeed the S-CSCF does not know it is anything other than a normal AS. It also has an interface towards the CAMEL (Customised Applications for Mobile Enhanced Logic) using the CAP (Camel Application Part) of standard SS7 signalling. CAMEL is basically a simple IN system for mobile – allowing roamed customers, for example, to dial short codes for their messages service and to be connected to their home mail box as opposed to an incomprehensible announcement in a foreign language that the number is not in service! In order to offer CAMEL services over the IMS the IM-SSF gateway is needed. For the IMS this is quite straightforward – let's suppose the user has an O2 mobile phone and has been used to dialling 901 to recover voice messages. When the user upgrades to an IMS handset it is still possible to use the same mail box and short code. A trigger is loaded in the S-CSCF so that a SIP URI of 901 is sent to the IM-SSF and from there, in the usual way, resolved by CAMEL to the actual number of the user's mail box (say 07702454567). When this is returned to the IM-SSF it simply replaces 901 with this number and sends it back to the S-CSCF – exactly as any IN system.

The Open Service Access Service Capability Server (OSA-SCS) is, in a similar manner, a gateway for OSA services to be executed on the IMS. Again, to the S-CSCF it looks exactly like any other SIP AS. OSA is essentially a Parlay interface – Parlay was developed by the Telco world to provide a standard interface for third party call control and creation. Essentially the OSA-SCS allows re-use of applications written in Parlay – it also allows a better degree of isolation between an operator's network and third parties and is expected to be used as the interface for third party services on IMS networks. As an example, in BT's 21C IMS test network there is a third party Parlay click-to-

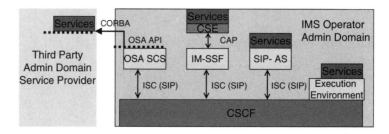

Figure 6-13. Application Servers for the IMS. (*Source*: Author. Reproduced by permission of © Dave Wisely, British Telecommunications plc.)

dial function. A user might be filling out an insurance application on line and "get stuck" or need help. A Parlay application can then (if it has access to their number) create a call between the user's phone and a call centre operator who can help with the application. The Parlay application to do this would contact an OSA-SCS and this would act as a back-to-back SIP user agent (a SIP proxy is not allowed to initiate dialogues!) and create a session between the two parties. (You can download the 21C SDK – Software Development kit at http://www.web21csdk.co.uk/ and create voice and SMS applications).

6.2.5 Presence

Presence is a familiar concept to anyone who has used MSN or Yahoo Messenger or the like. The idea of presence today is mostly to see who is on-line and whether they are available for a chat. In a multimedia IMS world it is possible to have a much richer presence in which users can publish information about:

- The various terminals they are connected through (e.g. PC, laptop, PDA);
- The network capabilities of the access networks they are connected through (e.g. Bandwidth, delay);
- The type of activity they are involved in (e.g. meeting);
- Their location (home, work, office, Lisa's bedroom);
- Their context – e.g. with visitors;
- Timed information – with visitors until noon.

Presence on the Internet works by means by means of a Presence server and a publish/subscribe model as shown in Figure 6.14. A central server holds information on the presence of "presentities" (i.e. those publishing information about themselves) – to which they publish updates. Watchers, seeking information on a user's presence subscribe to either recover presence information immediately or to be notified of any presence changes.

The IETF have created an implementation of presence using SIP by means of various extensions that add PUBLISH, SUBSCRIBE and NOTIFY methods to standard SIP. The actual presence information is carried in XML and can

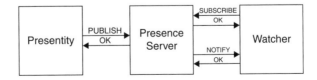

Figure 6-14. Presence model for Internet and IMS services. (Source: Author. Reproduced by permission of © Dave Wisely, British Telecommunications plc.)

include (with appropriate support) rich presence which includes, amongst other things:

- Activity – e.g. on the phone;
- Contact type – e.g. a PDA;
- Place type – e.g. library;
- Relationship – e.g. friend;
- Sphere – e.g. work.

In the IMS presence is naturally provided by a SIP AS. Now the IMS standards do not specify anything other than the standard operation of the IMS – the charging, session connection etc. functions that we have described in this chapter. There is no activity within 3GPP to look at standardising or working on the interoperability of common services – such as presence or Push to Talk. Instead these are being discussed in the Open Mobile Alliance (OMA – Ref. 3) – a group of mobile operators and vendors that work on the definition and interoperability of such services. The result of this activity is that it is now possible to buy a, more or less, standard presence AS that should interwork with other operator's presence offerings. Just about all current vendor IMS systems contain a presence server, for example. Presence information is clearly a building block for further services – such as answering machines, or chatting to any of my family who are on line etc.

3GPP defined the basic support for a presence architecture in Release 6. This includes a SIP AS Presence server, the addition of the SIP extensions for PUBLISH, NOTIFY and SUBSCRIBE and a Remote List Server (RLS). The remote list server allows users to subscribe and obtain information on a number of presentities without having the overhead of individually subscribing, and all the attendant SIP messages, travelling over the air interface. Instead the RLS subscribes on behalf of the user and aggregates the response to save signalling bandwidth. A new U_t interface – using X-CAP (XML Configuration Access Protocol) has been defined for the management of these lists.

As an example Figure 6.15 shows Tom (An AOL customer) subscribing to Lisa's (BT customer) presence information. This is forwarded in the normal way from P-CSCF (Yahoo), S-CSCF (Yahoo), I-CSCF(BT) to the BT S-CSCF that is serving Lisa. This has a trigger for the Subscribe method to forward these to the Presence server which can block or allow the request.

Figure 6.16 shows that when Lisa's presence changes the presence server is able to notify all the watchers – in this case Lisa is now busy in a meeting.

6.2.6 Push to Talk Over Cellular (PoC)

Another important service that is expected to be launched on the IMS is Push to Talk. This is a very familiar service to anyone old enough to remember CB radio

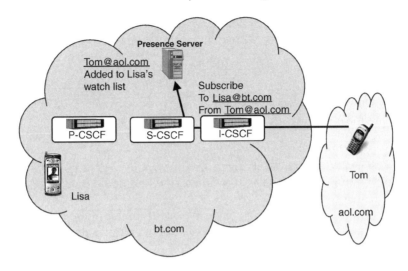

Figure 6-15. Example of the Presence service in the IMS. (Source: Author. Reproduced by permission of © Dave Wisely, British Telecommunications plc.)

("Come on Rubber Duck – 10/4") or those cheap kiddies walkie-talkies. The important point is that only one party can talk at a time with the need for users to say "roger" or "over" when they have finished speaking. Push to talk is much less demanding of radio systems than two-way voice – for example the delay can be much longer (500–800 ms compared to 100–200 ms for standard cellular voice) and only a single radio channel is needed for each conversation.

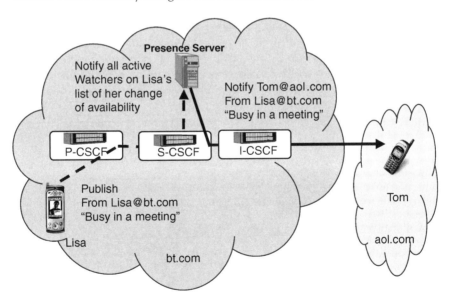

Figure 6-16. Notification of watchers. (Source: BT. Reproduced by permission of © British Telecommunications plc.)

For these reasons operators are likely to be able to deploy PoC on current, high delay, UMTS packet networks without the need to upgrade the delay performance. PoC has been a big success in the States – with proprietary solution like iDEN with 50 000+ subscribers in 2005. In Europe PoC has been held back by lack of standardisation and fear of cannibalising valuable SMS revenues. PoC can offer group conversations as well as one on ones. So, for example, the sales director can call his entire sales force and ask a question such as "What is the cost of a BT Fusion Phone on tariff 21"? Some of the salesmen will be busy, some won't know the answer but one will know and reply "£20", saving the director a lot of valuable time phoning them one by one until he got an answer. Again the OMA has been active in defining how PoC could be added to the IMS in an interoperable way (Figure 6.17).

In the IMS implementation the sales director in the example above could have set up a presence list of his salesmen and saleswomen and then only contact those that were online and available to receive calls by using the Presence service. The PoC server is used by all participants and handles both the signalling and media associated with the PoC session – even two way conversations go through the server – making it simple to add extra participants.

6.2.7 Roaming and Breakout to PSTN

6.2.7.1 IMS Roaming

As was explained earlier 3GPP voted for home control of IMS services – even if you have roamed to Australia from the UK your SIP signalling will travel back to a S-CSCF in England. Worse still, if the person you are trying to call has a home network in the US but is roaming in New Zealand, say, the signalling will go all the way to the US and then to New Zealand. This will, of course, add significantly to the delay in call set up and is the price to be had for complete control of network sessions. Figure 6.18 shows how complicated things get. At least Lisa and Tom can both get their own, customised services.

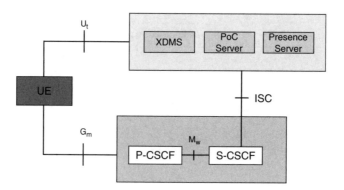

Figure 6-17. Push to talk over the IMS. (*Source*: Maria Qauvas. Reproduced by permission of © British Telecommunications plc.)

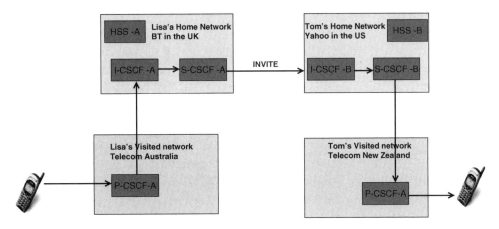

Figure 6-18.   Double roaming scenario. (Source: Author. Reproduced by permission of © Dave Wisely, British Telecommunications plc.)

So far we have always considered that the P-CSCF is located in the visited network –this is the long term plan and standard IMS architecture. However, where one operator wishes to deploy the IMS and uses IMS subscribers to roam to non IMS networks there is an alternative architecture – shown in Figure 6.19 – in which the P-CSCF and the R5 GGSN are located back in the home network and the user only makes use of the (non IMS) radio resources in the visited network.

6.2.7.2 Breakout to the PSTN

Obviously IMS services will be introduced gradually over a period of time and IMS terminal users will need a mechanism for making and receiving voice calls from the PSTN and other (non IMS) mobile networks. In this section we will see what happens when Lisa tries to make a call to her granny (who only has an old PSTN phone). Figure 6.20 shows the steps that take place.

After registration Lisa's terminal tries to make a call to Granny's number – 01473 645938 – this can easily be incorporated into the INVITE message as a tel URL:

Tel: ++44-1473-645675

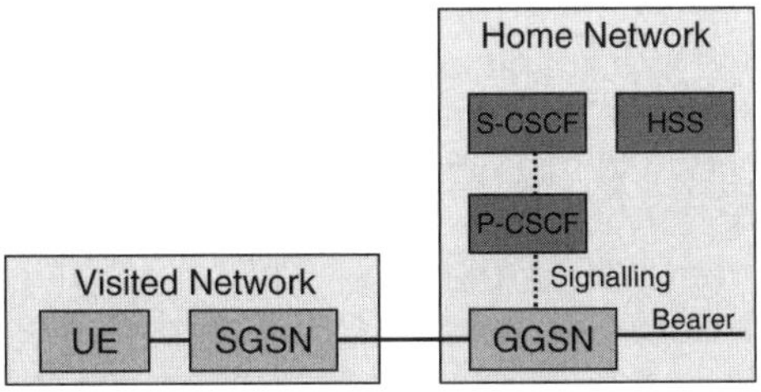

Figure 6-19.   GPRS-based IMS roaming. (Source: Author. Reproduced by permission of © Dave Wisely, British Telecommunications plc.)

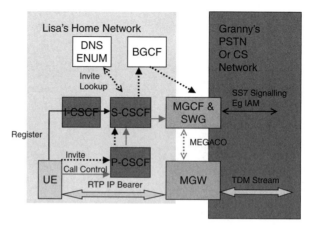

Figure 6-20. Breakout to PSTN voice from an IMS terminal. (*Source*: Author. Reproduced by permission of © Dave Wisely, British Telecommunications plc.)

When the S-CSCF receives this it tries to resolve it to a SIP URI using the DNS ENUM service (see below) – this basically allows the IMS to see if this tel URL is allocated to an IMS user who actually has a SIP URI and so can avoid leaving the packet network to complete the call. This is not the case for Granny, who has barely mastered turning the telly to Channel 4, so the S-CSCF needs to select a gateway to the PSTN to handle the signalling and media conversion from IP packet to SS7 TDM voice. The S-CSCF passes this job to a Breakout Gateway Control Function (BGCF – step 5). The BGCF then selects a suitable gateway point – possibly because it is close to the callee (based on the geographic number dialled) or it is the point of interconnect with another network.

There are two major functions that the Gateway needs to perform – conversion of the signalling from SIP to ISUP (SS7) and conversion of the media from an RTP IP packet stream to a 64 kbit/s TDM stream. The Media Gateway Control Function (MGCF) performs the signalling conversion and sends the ISUP messages over IP to a Signalling Gateway (SGW) that sends them over SS7 to a PSTN switch. For example the INVITE message is transformed into an Initial Address Message (ISUP) The MGCF also controls, by means of the MEGACO (Media Gateway Control) Protocol a Media Gateway (MGW) that actually connects the IP RTP stream and turns it into a 64 kbit/s TDM stream.

Calls to the IMS from the PSTN work in an analogous way (Figure 6.21) – each IMS user would be given a tel URL as well as one or more SIP URLs as part of their public user identities. These numbers will be linked to the IMS gateway of the issuing operator. When an ISUP IAM (Initial Address Message) is received by a MGCF it is converted into a SIP INVITE and passed to an I-CSCF which

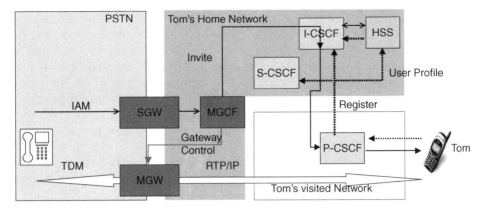

Figure 6-21. Call from the PSTN to an IMS user. (Source: Author. Reproduced by permission of © Dave Wisely, British Telecommunications plc.)

looks up the tel URL in the HSS to determine if the user is registered and if so at which S-CSCF. From the point of view of the IMS terminal this is just a normal SIP voice session request.

6.3 Convergence Today – Fusion

Somewhat ironically I am writing about BT Fusion on the day it is announced that BT is dropping the product! (Ref. 4). Nevertheless that does not mean that fixed-mobile convergence products are dead – far from it. In France FT had over 300 000 customers for their Unik phone in April 2007 (Ref. 5) that uses the same technology as Fusion and offers calls over WLAN when in range of a hot spot or access point. These products – and similar offerings around the world – rely on UMA – Unlicensed Mobile Access – a pre-IMS technology to allow voice calls to be made from a UMA phone over an IP backhaul and terminated in a normal mobile network. In this section we will look at how UMA works and then consider why Fusion has failed and what lessons that can give for future converged products. Then we will look beyond UMA to SIP/IP fixed-mobile voice solutions that are beginning to appear.

UMA phones combine two air interfaces into a single handset (that looks like a normal mobile phone) – GSM and WLAN or (previously) Bluetooth.[3] When in range of a home hub (or, sometimes, a public hot spot) the phone switches to WLAN for backhaul to the mobile network and the user is given an audio or visual indication that the switch has taken place.

Why would you want one? Well most operators offer cheaper calls when on WLAN – BT offered three minutes on WLAN for every 1 minute of monthly call

---

[3] Originally it used Bluetooth to connect to the home hub but now all UMA handsets use WLAN and I will not mention Bluetooth again.

allowance used. Secondly, indoor coverage is still far from ideal the power level at the phone falls 10 dB (i.e. by ×10) when you move indoors and sit by a window and by 20 dB (i.e. by ×100) when you are well inside. We noted this on our discussion of femtocells in Chapter 4. Those are the user benefits – so what is in it for the operators? Well, firstly, they can start to take traffic from fixed operators – why do you need a fixed phone if your mobile offers cheap calls indoors anyway? Secondly, it saves backhaul costs to route the calls through DSL as opposed to expensive E1 links in the GSM radio network. Finally, it is a further step in the direction of triple or quad plays and a stepping stone to controlling all of the telecom/TV services that users take (tomorrow the universe!). In fact some mobile operators have been offering lower cost calls in the user's home radio cell – e.g. O_2 are offering a successful home zone service in Germany called Genion. Mobile operators, however, have been having problems with these "home zones" extending over considerable areas (even in dense urban environments a GSM base-station might reach 500m). Users with free call tariffs have not been unknown to take the phone down the pub and pass it around. These services have been very popular with users – who, it seems, are only restrained from making even more and longer mobile calls by worries over cost! Operators are keen to exploit the appetite of users for these home zone tariffs – if only to differentiate themselves from other operators – whilst limiting their use: hence the attraction of UMA.

UMA is, in fact, a simple technology. Figure 6.22 shows how a UMA phone connects to a mobile network over a WLAN interface. GSM signalling (MAP) and time division multiplex (TDM) voice is encapsulated in a secure IP tunnel

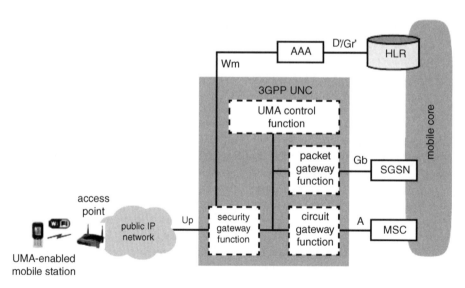

Figure 6-22. Unlicensed Mobile Access (UMA) – network architecture for WLAN call – GSM calls go through the mobile network as normal. Courtesy of the BTTJ. (Source: BTTJ. Reproduced by permission of © British Telecommunications plc.)

from the handset to a UMA network controller (UNC). The UNC recovers the GSM signalling and TDM voice stream and then presents it to the mobile network exactly as if it had come from a normal GSM base-station controller (BSC). The great advantage of UMA is that it offers seamless handover between GSM and WLAN and users just have a single mobile number. We will see that SIP/WLAN phones don't always achieve this.

There have been a number of issues with UMA that have tended to hold it back. Firstly, the biggest issue has been handsets. It is often reckoned that to launch a new service requires at least 10 new handsets – because you are targeting so many different market segments and users really care about handsets (young, old, poor, businessman, student, librarian, people with sticky-up hair...). UMA handsets have proved difficult and expensive for manufacturers to develop and, consequently, there have been only one or two available at the launch of many UMA services. Now more are on the market but there is still a limited choice and the very latest handsets are not UMA-enabled. Secondly, the battery life of UMA handsets – despite much tweaking of the WLAN – means they need very frequent recharging (or many users turned the WLAN off – which defeated the point of having the phone in the first place!). Thirdly, the coverage of the access points was found to be very limited – much less than their range with WLAN laptops, say. This was mainly because the phones were very conservative in initiating a handover as soon as they saw the WLAN signal dropping away – to ensure the handover (taking 1–2 sec) was complete before the user had walked out of range of the access point. Finally, some tariffs were not attractively set. Users, it turns out, are often on the wrong tariff and either under use their call allowance (and should move to a cheaper tariff) or go over their limit and incur usurious costs. They are just not that price sensitive – so the home calls need to be quite cheap to attract attention.

One of the more promising areas for converged handsets is in the enterprise space. I am, I suppose, typical of the sort of enterprise user that financial controllers hate. When I am sitting at my desk and want to call somebody I just get out my mobile phone, find the contact, and call them. Sitting in front of me is a perfectly good fixed phone. Moreover this fixed phone costs my employers nothing for me to ring internally, but I don't use it: I don't use it because it hasn't got my contact list on it, I can't walk around with it and, most importantly of all, my mobile is also paid for by BT so cost is not a factor for me! There is also a (I suspect apocryphal) story that a well-known company found its employees had sent 76 000 picture messages in a year despite not issuing a single one of them with picture-messaging capable phones! (they took the SIMs out!).

All of these things point to a very high potential demand for converged phones that use WLAN within the enterprise and can use hot spots when roaming. BT has developed two solutions (Figure 6.23) both of which use SIP/IP rather than UMA. The first variant (Site Fusion) simply offers a WLAN VoIP client on a mobile handset. There is no integration between the GSM and WLAN – with no handover and separate numbers in use. Nevertheless it is reported that the biggest source of growth at BT Openzone hot spots is VoIP. The second variant (Enterprise Fusion Net) offers more functionality. A single

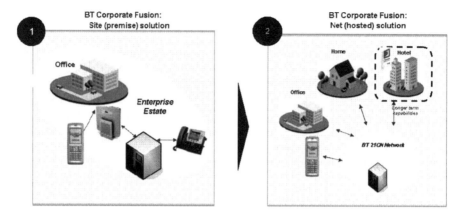

Figure 6-23. BT's Enterprise Fusion. Site (left) and Net (right) versions. (Source: BT. Reproduced by permission of © British Telecommunications plc.)

phone number (07777 XXXXXX – belonging to BT) is used and call handover – between GSM and VoIP is handled by a system known as Voice Call Continuity (VCC – this is explained in the next section) – which uses the IMS platform and IN connections to the Vodafone GSM network. Early tests of the system show that it requires upgrades to the enterprise WLAN system and that the battery life of the handsets when operating in WLAN mode can be only a day. However, given there is a major drive by many enterprises to get a grip on mobile costs, the lack of handsets is less of an issue when they are dolled out by the company, and if technical set-up is handled by a dedicated team then this kind of product looks to have a brighter future than UMA-based solutions.

6.4 Voice Call Continuity

UMA is really a 2G solution for circuit voice between cellular and WLAN and has nothing much to do with the IMS. But you can imagine, as the IMS is rolled out, that there will be occasions when there is a need to handover between circuit (CS) and IMS-controlled packet (PS) domains. Examples might be where an operator has rolled out IMS for WLAN VoIP and multimedia services – so all voice calls started on WLAN are VoIP. However, when the user walks out of range of the WLAN it is possible that VoIP is not supported on the available cellular network (true today where 3G coverage is very patchy and only the latest HSPA supports VoIP). There might also be occasions where operators want to switch CS voice calls to IMS-controlled VoIP. A good example might be that a user starts a bog-standard voice call and then invokes advanced features that only an IMS can supply. The operator transfers the call from the CS to the IMS and then invokes the advanced features (at extra cost to the user no doubt).

3GPP have come up with a framework that transfers calls between the CS and IMS –controlled PS called Voice Call Continuity (VCC). Basically this is an application that runs in the user's home IMS. VCC subscribers then find that all their call control (whether roaming or at home, when originating on CS or PS) is anchored in this application. Figure 6.24 shows a user terminal with both IMS and CS application stacks. If the user originates a call from the CS domain the bearer is converted to IP at a media gateway and the signalling converted to SIP at a signalling gateway and then goes to the S-CSCF. In the case of a VCC subscriber the SIP messages pass to a special Application Server (AS) called the Call Continuity Control Function (CCCF). This only comes into play when a call is transferred. For calls originating on the IMS the bearer (IP) goes straight to the remote end but again the CCCF puts itself in the path of the SIP signalling – so that a transfer to the CS domain is possible.

There are many possible combinations of domains to transfer from and to and they get a bit (alright very!) tedious so I will use just one example (Figure 6.25). This is a real example in that it is the BT Enterprise Fusion case of a user with IMS VoIP on WLAN needing to transfer the call to the CS domain as the user walks

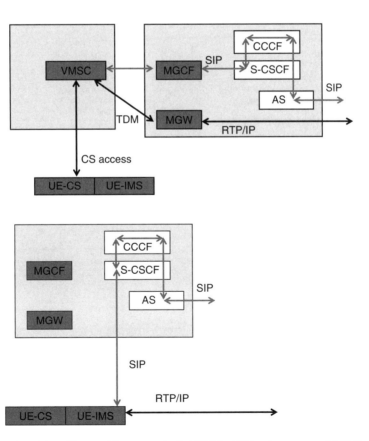

Figure 6-24. Signalling and bearer paths in VCC for CS (top) access and IMS (bottom) access. (Source: Author. Reproduced by permission of © Dave Wisely, British Telecommunications plc.)

out of range. This is a real application running on a real IMS and is working in live systems.

In this example a call (VoIP) is in place between the mobile in WiFi mode and an IP phone. When the handset detects that the WiFi signal strength is fading it determines that a GSM handoff is needed. The handset then establishes a GSM call using a handoff routing number (which belongs to the MGC/MGW which was downloaded to the handset during the registration process). The MGC converts the IAM (Initial Address Message) into a SIP INVITE which is routed to the S-CSCF. The S-CSCF triggers on the handoff routing number to pass the INVITE to the CCCF. The CCCF uses the INVITE from field – which holds the mobile MSISDN number to correlate the handoff request and the existing call. The CCCF then instructs the MGC/MGW to switch the call from the WiFi to GSM network by sending a RE-INVITE with the new session SDP via the S-CSCF. Once the MGC/MGW accepts the change of session SDP, the CCCF then connects the new session over GSM with an OK to the MGC/MGW and then disconnects the WiFi connection by sending BYE to the UE. If you want to read

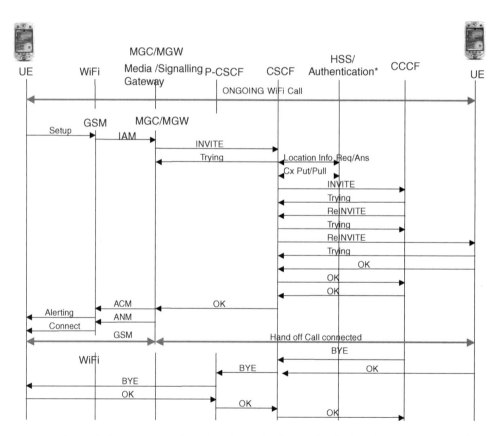

Figure 6-25. WiFi to GSM handover for Enterprise Fusion (BT). (Source: Author. Reproduced by permission of © Dave Wisely, British Telecommunications plc.)

about the full standard way of transferring calls and about the (complex cases) of roaming then download TS 23.206 V1.1.0 (2006-7) from www.3gpp.org.

VCC was completed in Release 7 but there are still a number of new areas being considered: The multimedia session continuity (MMSC) work in 3GPP (Rel-8) introduces significant enhancements to the previous VCC work in Rel-7. Unlike VCC, which supports only voice and for domain transfer between CS and PS only, MMSC provides solutions that support:

- CS-to-PS or PS-to-PS domain transfer;

- Other media components (voice, video, data);

- User's ability to add, remove, split, retrieve, or combine media components in one session;

- The media 'transfer' to other UE.

The MMSC study was completed in 3GPP (Feb. 2008) and the focus now is to progress to the specification phase – IMS Service Continuity.

6.5 Media Independent Handover

If VCC is about handover between the circuit and packet worlds then what about handover between the different IP packet networks? Between WLAN and WiMAX, for example. Well IP handover is a problem that has been kicking around a long time, see Ref. 6 for a very tedious list of solutions that have been worked out over the past decade to the problem of handing over between IP networks. In summary – to save you from the fate that befell me (banging my head on the table when nodding off) whilst perusing the above mentioned guide to IP handover – there are solutions at different layers. A layer 2 solution for handover is useful for intra technology handover within a single IP subnet (i.e. all your WLANs are on a single IP subnet then use a WLAN layer 2 handover protocol – as described in the WLAN chapter). This is often called a horizontal handover. If the handover is between technologies (Inter technology or vertical handover) then either a layer 3 or higher solution is needed. The basic problem is that the terminal's IP address changes and network applications that are running don't always survive this change.[4] Mobile IP (MIP) is the best known solution at the IP layer – in its simplest form this tunnels IP packets from a fixed point (and IP address) in the network (the home agent) to the mobile host. When the mobile host moves it notifies the home agent and the packets are re-directed (see Ref. 7 for a good explanation). The problem with MIP is that it takes 10–30 secs to achieve a handover and so is not suitable for real-time services. SIP can

---

[4] Smart FTP will – as will some applications like web browsing and email on a Windows XP machine – try it! – connect via Ethernet, start browsing and then pull out the lead. If you are in range of a WLAN and have already configured the security details then XP will sort out the handover and the sessions will continue. Typically this will take 20 sec to 1 minute to happen.

also be used for handovers involving a change of IP address. This is actually one of the research topics of my own convergence team's research – so, for fear of blowing my own trumpet (and covering the readers with phlegm) – I will only summarise what we have done in this area and leave interested readers to read the details (Ref. 8). With this handover the SIP user agent is able to interface with a connection manager and get access to the IP address of the current network. A change of IP address can then be sent in a RE-INVITE SIP message to the correspondent host. SIP applications using this method have to be re-written so that they are able to send both SIP and RTP packets to the new address (Figure 6.26 shows the handover signalling).

The basic problem with MIP, SIP and, indeed, application layer handover is that it takes too long for real-time services. Figure 6.27 shows the delays for a SIP-based handover between WLAN and WiMAX networks. You can see the handover takes about 10 sec – compared to (say) 30–100 msec for a typical layer 2 handover in mobile or with WLAN to WLAN.

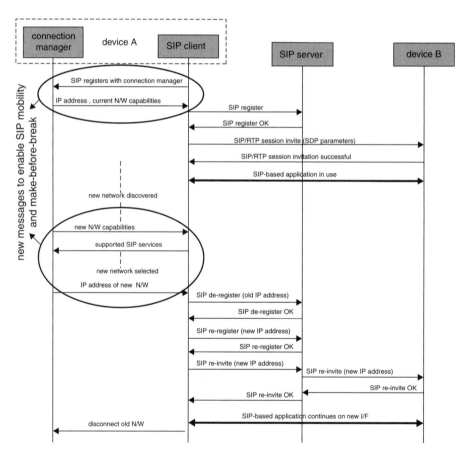

Figure 6-26. SIP-based handover (see Ref. 8 for more details). (Source: Author. Reproduced by permission of © Dave Wisely, British Telecommunications plc.)

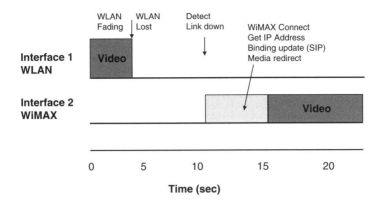

Figure 6-27. Handover delay WLAN to WiMAX – break before make. (*Source*: Author. Reproduced by permission of © Dave Wisely, British Telecommunications plc.)

The delay is far too long for voice or video services – 10 sec or so. The basic problem is that the only way for MIP, SIP or application layer solutions to bring up a new network and switch to it is to wait for the old network to lose connectivity. In other words, pure layer 3+ solution have no knowledge that the old network is losing signal strength and simply have to wait for connectivity to disappear and the connection manager (whether Windows XP or whatever) to bring up a new network – Figure 6.27 shows what happens with an XP terminal. This sort of handover is called break-before-make – essentially the old network connection is broken before new one is made.

What is really needed, for seamless handover, is make-before-break. This is where the connection manager (or whatever software is controlling networks) has an algorithm for determining whether a link is failing and brings up a new interface in advance of the old link going down – so that much of the delay in the handover (getting the new IP address etc.) is avoided. Make-before-break handover, however, is something that needs cooperation from layer 2 – that is where the network drivers are located and where information is available on signal strengths and available networks.

Traditionally, this coupling between layer 2 and higher layers has been done in a proprietary way. For example, the UMA phones that handover between WLAN and GSM use WLAN signal strength as a metric for handover and simply initiate handover whenever it falls below a certain level – the handover takes several seconds (even though the GSM interface is up and operating all the time) so this is set fairly conservatively to allow the handover (so you have to be close to the hub to get the cheap calls!). Previously it has been possible to speed up MIP by linking the client to a Linux WLAN card driver – the Linux driver offering access to the signal strength and other parameters not possible under Windows, for example (see Ref. 9).

In order to standardise the interface between layer 3+ mobility solutions (which can include: MIP, SIP and UMA) the IEEE 802.21 working group has

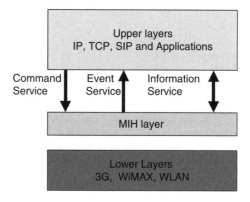

Figure 6-28. 802.21 Media Independent Handover. (*Source*: Author. Reproduced by permission of © Dave Wisely, British Telecommunications plc.)

introduced Media Independent Handover. The standard introduces three services (Figure 6.28):

• Event Service;

• Command Service;

• Information Service.

The Event Service is an API that provides standard messages – such as prediction that the link (e.g. a WLAN connection) is going down and then confirmation (link down) that the link is lost. The Command Service provides a standard way of interfacing to the network drivers to bring up interfaces and initiate handovers. The Information Service can provide information about available networks (channel, QoS, authentication needed etc.). It can do this for networks that can't be directly sensed (e.g. because you have the WLAN radio off to save power say). Contextual clues (e.g. location or cell-id) are sent to an Information Service server and information on available networks is returned.

Of course this sort of handover technology is mostly useful for WLANs where the coverage is patchy. For WLANs the signal strength can vary very rapidly over a short distance – but this doesn't always mean that you are reaching the edge of a cell – it might just be that you have moved behind a metal filing cabinet say. One area of active research has been to take the signal strength from a WLAN card and build an algorithm that turns it into a LINK-GOING-DOWN prediction for the Event Service that is both accurate and sufficiently in advance of the WLAN failing completely. Figure 6.29 shows the improvement that can be made by using one such algorithm (Ref. 10) together with make before break handover. This algorithm reports to be 96% accurate at predicting the WLAN connection failing at least 1.1 sec in advance. It can be seen that with this optimisation that make before break handover is now possible. As can be seen in Figure 6.29 this allows seamless handover (i.e. loss less and sub-50 ms

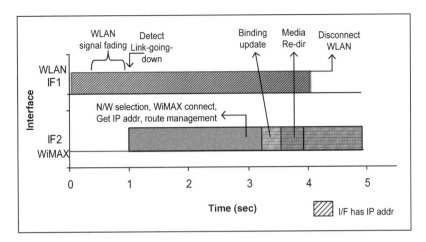

Figure 6-29. Make-before-break handover with 802.21 Media Independent Handover and smart trigger for WLAN network loss. (Source: Author. Reproduced by permission of © Dave Wisely, British Telecommunications plc.)

break). With audio and video applications running there is no obvious break when handover takes place.

The Information Service is controversial because it is the key element of network discovery and selection. Let me illustrate. I was sitting in the Gare Du Midi in Brussels (South station for you Francophobes!) – and wanted to read my email. So I fired up the old laptop – it didn't connect to anything – so I looked at the list of WLANs. Six WLAN networks were available. Mostly with obscure names that gave no real insight into how to log on: which ones were partners to BT Openzone? In the end I tried them all! And found number 3 on the list was unsecured and free – and so I used that – although I am not sure I would have sent my credit card details over it![5] The point being that discovering networks, in a world of multiple WLAN and (in the future) WiMAX networks, is not that easy. This is where the Information Service (IS) comes in – from a location (say) you can use 3G or GPRS to find out about WLAN and WiMAX networks. The IS provides information about available networks such as:

- Network type;
- Frequency/channel;
- Owner;
- QoS support;
- Emergency service support;
- Price.

---

[5] I might to a site that supports https – the secure sockets layer that is included in most popular browsers (including IE).

Of course this is very controversial! Who will provide the IS? Operators like BT? Regulators? Google? The last thing operators want is for users to sit in a café and compare network prices – having gone to all that trouble of signing people up to expensive contracts. This really starts to impinge on the commercial battleground of the future – which I am postponing for the final chapter. However, the actual selection of network is far from simple. If you are a poor student then you might only want low cost networks, if you are cycling along there doesn't seem much point connecting to a WLAN hot spot only to cycle through it in 2 sec! And WLANs are not always cheaper – what if you have a 300 min/month call allowance on 3G and you have only used 200 min by the last day of the month – then 3G is free for you.

The actual handover decision is made by a connection manager. Connection managers are everywhere – there are explicit ones in XP and the MAC OS (Kitten or whatever version they are up to!) and UMA phones have a simple one (find a WLAN with this SSID – attach. Lose WLAN signal strength – handover to GSM). The trouble is that with richer commercial models than UMA (different terminal, GSM and WLAN suppliers) and richer services (Video, web. . .) the whole problem becomes much more complicated.

In fact research is only just beginning in this area – Figure 6.30 shows the configuration my team (in conjunction with Nokia and Intel – Ref. 8) is working on. We have separated the connection manager with interfaces from the 802.21 services (Event, Command and Information). Our SIP client has an interface to the connection manager – allowing it to get information about network changes (such as the IP address of the new interface) and to inform the

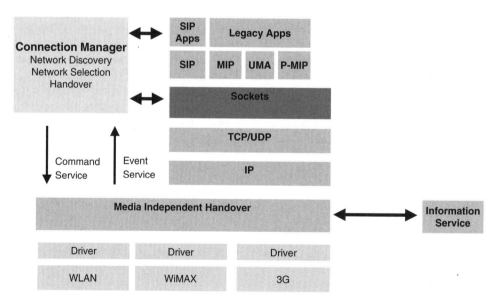

Figure 6-30. Outline client architecture for a Media Independent Handover capable terminal (author's lab). (Source: Author. Reproduced by permission of © Dave Wisely, British Telecommunications plc.)

Figure 6-31. Screen shot of our SIP voice/video/gaming client (author's lab) – who is this man? Don't worry – "he's from Barcelona" (Source: Author. Reproduced by permission of © Dave Wisely, British Telecommunications plc.)

connection manager of the terminals capabilities (e.g. codecs) to consider in making the network selection. We have also made MIP work with make-before-break handover.

Finally, we have written new SIP applications – for voice and video – that take advantage of the information about networks available from the connection manager to adapt to changes of bandwidth or QoS (by means of codec switching). So far we have been able to switch seamlessly between WLAN, WiMAX and 3G – you can see a screen shot in Figure 6.31. To summarise, and avoid going on about our own work, the aim of the research is to figure out whether there are simple practical rules that work well for a connection manager or whether significant performance gains can be made by referring handover to the network – which has access to more knowledge about things like QoS and network congestion. To date we are still working on it!

6.6 Next Generation Networks

Next Generation Networks sound like something out of Star Trek but in fact NGNs are just a way of extending the IMS to incorporate the fixed DSL IP network. One of the best examples of an NGN is BT's 21C network. In days of yore BT had a complicated network with a mixture of different networks: SDH, PHD,[6] ATM, IP, leased lines, private networks and the PSTN. The

---

[6] If you don't know what SDH and, especially, PDH are then don't ask – you'll regret it – trust me!

number of connectivity products was literally hundreds and it was getting very expensive to run just as competitors (e.g. cable) were coming on the scene. It was also next to impossible to introduce new services – requiring endless IN overlays and lots of special coding (costly) to get even, by IP standards, seemingly trivial services to work (e.g. call minder). In the end it was decided to move to an all-IP network – a Next Generation Network. Figure 6.32 shows the "before" and "after" pictures.

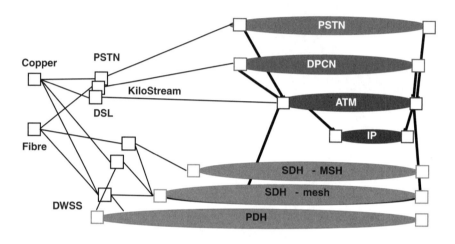

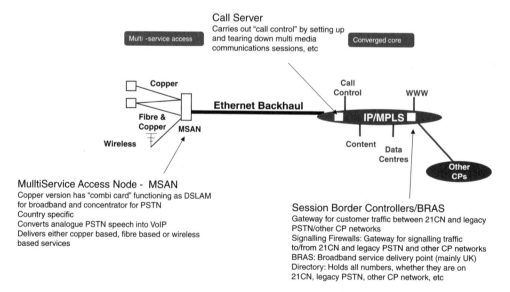

Figure 6-32.   BT's UK network – network in 1999 and in 2009 after 21C transformation. (*Source*: Author. Reproduced by permission of © Dave Wisely, British Telecommunications plc.)

There was a lot of debate about 21C – particularly what "all-IP" should mean for the PSTN: should a completely new end-to-end VoIP system be introduced – meaning SIP signalling and IMS-type services – or the existing PSTN be emulated – with traditional phones and only current PSTN services. In the end it was decided that the network would be IP from the MSAN (Multi Service Access Node– the local exchange in effect). However, a copper line would continue to carry normal PSTN signalling and TDM voice data from the home to the MSAN. At the MSAN the PSTN signalling and voice would be converted to IP – i.e. the TDM voice stream would be packetised and the signalling converted to SS7 over IP. The actual call control is provided by call servers at so-called Metro-nodes. Call servers are essentially the control plane part of exchanges – so this architecture is very like that of 3GPP's R4 mobile set up as described in Chapter 2. The upshot is that – for the PSTN – you will not notice any difference – the same phones will plug in to the same sockets and still use E164 number dialling and offer emergency calls, powered back-up and guaranteed QoS. For BT the new PSTN offers a much simpler network, with fewer nodes, fewer protocols and less expensive legacy kit. In addition new PSTN services are easier to introduce – as we will see it is possible to link the new call servers to more advanced IP services. The other advantage of 21C is that changes of service, such as DSL upgrades, activations and second lines, can be achieved in near real-time. BT will spend £10B between now and 2010 or so, rolling out 21C in the UK with expected saving of £1B/year. BT have also just (Q2 2008) announced £1.5 billion investment into fibre to the premises (i.e. all the way to the house/business) or to the cabinet (very close to a small number of houses) to extend the IP reach of 21C.

21C, however, is about much more than replacing the PSTN. 21C is also about new IP services over broadband to both enterprise and consumers. As we have noted earlier, being a DSL provider is very different to providing data services (like SMS, ringtone, music download etc.) over mobile. ISPs provide just a data transport capacity – 8 Mbit/s max speed (really 3–4Mbit/s) with a monthly cap of 5Gbyte for something like £15 a month in the UK. Prices have been falling so much and speeds increasing so much that if you calculate the cost/month/Mbit/s you will find that dial up used to be £12/month for 56 bit/s (£240/month/Mbit/s) – three years ago it was £26/month for 1Mbit/s (£26/month/Mbit/s), last year you could get 8 Mbit/s for £24/month (£3/month/Mbit/s). Today Virgin is offering 24Mit/s for £15/month (£0.63/month/Mbit/s). You can see where this is going. . . .£0 month/Mbit/s. . . internet connectivity for free.[7] Even if you don't believe that then being an ISP is

---

[7] I have long discussions with colleagues on whether broadband will be free. If you are not convinced then go to MacDonald's – in the UK they offer free WLAN broadband. Sky TV offer "Free" broadband if you take their TV service, Talk Talk offer free broadband if you take their phone service. You can get broadband on the move from Vodafone for £15/month – HSDPA, 1Mbit/s real throughput, 3Gbyte monthly limit. Of course DSL is not free – somebody has to dig up the roads and fix the electronics when it breaks and pay rental on the building. However, this is getting down to the £5/month level that other revenue (e.g. advertising) can easily offset.

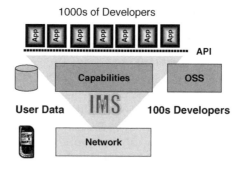

Figure 6-33. *BT's 21C Service creation for IP services.* (*Source*: Author. Reproduced by permission of © Dave Wisely, British Telecommunications plc.)

very competitive and puts you at the bottom of the Internet business stack. People do amazing things on the Internet: buy books, book tickets, IM, VoIP, video calling, Facebook, IPTV and the list goes on. But the ISPs don't get any of this revenue – it all goes to the likes of Google, Skype, Amazon and Visa. 21C is BT's answer to capturing some of the value of these services.

Essentially the non PSTN part of 21C consists of an IMS (Figure 6.33) and other "reusable capabilities", such as messaging and presence, that can be put together to offer novel IP services over DSL and WLAN (with linkages to the call servers of the PSTN possible). The idea is that not only BT applications will be offered but there will be an interface to allow third parties to create services – as mentioned earlier there is a 21C SDK (Software Development Kit) that you can download at http://www.web21csdk.co.uk/. This reflects, of course, the reality that telcos have an appalling record of creating useful IP services compared to one man and a dog (sorry Larry but Sergi is the good looking one!) in a shed in California.

Currently there are a number of IP services running on the 21C network including Broadband voice – a VoIP service that offers a SIP-based VoIP client with video calling and buddy lists etc. Another application built on the 21C platform is Go Messenger – a client for the PSP (Play Station Portable you old fogey!) – this allows messages and video/voice calls to be sent between PSPs using WLANs such as home hubs and hot spots.

Some of these new services require QoS support – not only services like voice and video but also services that the operator is keen to differentiate from similar offerings from Internet based companies. For example there are lots of companies offering DVD download (for sale or rental) – so if I, as an ISP, want to offer a similar service I might try and be cheaper than Internet offerings (unlikely) or I might offer faster download times – or something novel like watch 10 minutes of the DVD for free – using streaming and, if you like it, buy the right to watch the rest (which I will stream to you). Maybe, however, the standard DSL offering doesn't have sufficient QoS to support this service (in the busy hour the throughput is 400 kbit/s max say – too low for 2Mbit/s streaming of a standard DVD). If, however, there was a QoS mechanism that could guarantee 2Mbit/s to my customers then I could launch the service and the Internet-based DVD rental

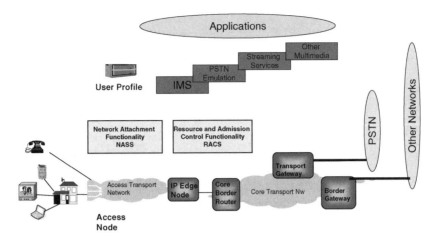

Figure 6-34. ETSI TiSPAN NGN Release1. (Source: Author. Reproduced by permission of © Dave Wisely, British Telecommunications plc.)

companies could not compete. Also I want my VoIP service to work reliably – so people think it is a real alternative to mobile or PSTN and I can develop value-added services using the IMS (smart call diversion, video calling...).

There is a general architecture for QoS in DSL networks that is part of ETSI's standardisation of Next Generation Networks. Figure 6.34 shows the overall scheme of ETTIs Release 1 for NGN. If you look closely you will see that it is mostly based on the IMS architecture and, indeed, standardisation of the NGN has now passed to 3GPP from ETSI.

The similarities to a 3GPP R5 architecture are striking:

- Layer 2 tunnel to edge router (IMS: GTP tunnel to SGSN);

- IP core to Border Gateway (IMS: IP core to GGSN);

- Media Gateway to PSTN (IMS: Media Gateway to PSTN);

- IBCF (Interconnection Border Control Function) (IMS: Gateway Control Function);

- RACS – Resource and Admission Control Subsystem – (IMS: Policy Decision Function);

- Applications built on call/session control (IMS: Application Servers).

You can also see how PSTN emulation fits into the NGN – to sit alongside the IMS as another call/service control function. Coming to the RACS (Resource and Admission Control Subsystem) – this is how QoS is delivered in an NGN. Unfortunately this is different from the R5 QoS scheme and is one of the proposed harmonisation items for the evolved IMS now that 3GPP is responsible for all IMS/NGN standardisation. Figure 6.35 shows a very simple example of three services wanting QoS in an NGN.

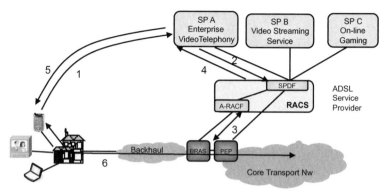

1.   Setup videotelephony conference to team
2.   Request resources for audio and video flows specifying characteristics
3.   Make policy-based and resource availability-based admission decision and enforce
      corresponding policies in the network
4.   RACS sends acknowledgement of successful reservation to Application
5.   Application acknowledges successful videotelephony session to user
6.   Videotelephony session starts

Figure 6-35.   Typical application of RACS in an NGN.

An application function is contacted either by an IMS with a service request or by applications running on a terminal or home PC to request resources. This request is referred to as the Service Policy Decision Function (SPDF) – this is responsible for ensuring that the request for QoS is valid – e.g. if I am asking for QoS on video calling that I have actually subscribed to that package. Or perhaps QoS for DVD streaming is only supported when the service is mediated through the IMS. These are examples of the policy-based admissions of NGN QoS. The SPDF then contacts the Access Resource and Admission Control Function (A-RACS) – which is responsible for admission control – to see if resources are available and, if so, reserve them for this application. The A-RACS is also responsible for policing user traffic on layers 2 (ATM) and 3 (IP). For DSL networks the A-RACS function is actually implemented in the BRAS (said Bee Ras–and nothing to do with underwear) – the broadband access routers in the main exchanges that terminate the ATM VCs from the home hubs. The SPDF also contacts the border gateway function (BGF) this enforces policies and performs firewall functionality under control of the SPDF such as:

• Gate control – security;

• Packet marking;

• VPN support;

• Hosted NAT traversal;

• Policing of down/uplink traffic;

• Usage metering.

So, in summary, it is very like the IMS and mobile packet network in operation. NGN R1 still leaves a number of issues that will now be dealt with in 3GPP:

- Policy decision and enforcement points in the upstream direction;

- Admission control over core and interconnect transport networks;

- Extend RACS to support other IP Connectivity Access Network types (e.g. UMTS/GPRS);

- Achieve true end-to-end QoS by considering the Customer Network segment.

6.7 The IMS in Context Today

Today there are very few IMS systems actually in operation and in this concluding section we will look hard at the reasons for this and at some of the alternative solutions for service creation in IP networks. The first point worth making is that the IMS was conceived in the late 1990s – when the Internet was in its infancy – so it is, by the standards of the Internet, a bit old hat. The big idea of the IMS was to capture the value of all those Internet services that were current 10 years ago. The IMS was designed by telecoms engineers to make the Internet look like a traditional telco business, so was heavily based on guaranteed quality, voice and "Rolls-Royce" engineering. Telcos also wanted it to be backwards compatible with existing systems and models. So problem number one is that the IMS does not follow the Internet service creation/business model. Problem number two is that its services are several years out of date and problem number three is that it is a complex platform that requires security, QoS and a whole host of things to be in place for it to operate successfully.

The Internet has changed beyond all recognition in 10 years. The key applications are (today at least) all about Facebook, P2P file sharing, social networking, email, IM, Search etc. The key players are not telcos but Internet start ups– Google, Amazon, latsminute.com etc. Services in this new WEB 2.0, as it is called, are created in a totally different way than the IMS envisaged. The other issue is that the key voice/video/messaging/presence applications of the IMS have been around for years on the Internet for free. If you have not signed up for MSN, Yahoo, Google or one of the other email/voice/video and messaging services then, quite frankly, where have you been for the last 15 years? The serious point is that nobody is signing up with the Mobile Operator's own (unknown) services – the land grab phase is over. No, mobile operators are having to partner with Internet giants like Microsoft and Google. In a landscape of rapidly changing services and rapidly changing business models flexibility is the key. The Internet, with its open, IP architecture has been a fertile hotbed of innovation and evolution that the more rigid telco world has sometimes struggled to keep up with.

With the very slow rise of mobile data on 3G networks mobile operators have not felt the need for an IMS. Technically, as we have seen, mobile 3 and 3.5G networks have not had either the coverage, delay, bandwidth or all three to support VoIP. Even if they had what would be the point of moving standard voice from the CS to the PS domain when you have so much spare CS capacity and where, call for call, it is much cheaper.

The other issue for the IMS is cost. You can buy an IMS from a leading telco vendor – they will charge a very significant sum for the hardware. Then you will need them to set it up, there will be on going maintenance and support fees. When you connect any users you will need a licence for each client. Now suppose you wanted to launch a VoIP service over the fixed Internet (DSL and WLAN say) – maybe some messaging, video calling, presence lists and so on for good measure. You are putting together the hardware options for a business case. You could buy an IMS – but to operate the full system you would need an HSS and to issue every user with a SIM card (and a SIM card reader for their use). The SIMs alone would cost £5 each and the same for the reader – that would destroy your business model as there is so much competition, barriers to entry are low and many established VoIP providers allow you to sign up for free – so you would have to do the same and all those people who used the SIM and reader once would cost you £10. Next you would look at the IMS cost versus alternatives. Well there are plenty of open-source "free" SIP servers available – that would run on servers you could pick up at PC World for £1k or so. SIP user agents are also available – they would just need a small amount of customisation with a fancy log-in screen maybe. You get the picture – the IMS is only really worth all the costs for big brand companies who will exploit the security, QoS etc. – e.g. by using the fact that you are authenticated to allow on-line purchases without needing to type in a credit card number – for example. Telcos will also be able to offer important interconnect services, enterprise integration with local PBXs, international connections between VoIP sites with QoS and so on. But it is a big investment and so far, companies like BT have shied away from the cost of rolling a fully fledged IMS and stuck to an IMS-lite – essentially a commercial grade SIP proxy server. In addition BT has used a non IMS technology for messaging – XMPP – and non NGN QoS kludges to get the service working. This has reduced the risks, lowered initial investment and allowed greater flexibility. It is also sometimes argued that IMS systems are not truly scalable – especially the SIP presence model. Indeed all the major presence/VoIP/messaging suppliers (Skype, Microsoft, Google and Yahoo) do not use SIP-SIMPLE but proprietary software. This is a problem if you buy an IMS as it will not interwork easily and any clever IMS service you develop can't be easily reproduced.

At the moment the jury is still out on the IMS – mobile operators don't yet have a low cost VoIP capability and, without VoIP, there seems only slight motivation IMS. They have also chosen to partner with existing Internet operators of IMS-like services – locking in to proprietary systems. Fixed operators need to capture value from Internet services over and above bit transport and have been more pro-active in deploying SIP-based VoIP services. However, the full IMS system is too expensive and "gold-plated" in an Internet age of P2P, Facebook, Skype and so on. So far they have stuck to basic, pre-standard, SIP proxy servers and are waiting to see how the Internet moves. In my view we are unlikely to see a full IMS system deployed before WiMAX and LTE are rolled out. These are all-IP networks that must use VoIP but are too expensive to be just sold as mobile broadband with users free to use Skype, Microsoft Messenger etc. for service creation. No these are fully-fledged mobile networks that will need to offer VoIP with quality guarantees, emergency calls, business services and so on that will command the premium price needed to support an IMS system.

Is convergence 4G? I think it is much closer to the 4G vision that was current in the last years of the old millennium when the Internet was just exploding and 3G was still being refined. The idea that each new mobile generation has to be a new air interface, handsets and network is running out of steam – with 3.5G (HSPA) blurring into 3.9G (LTE) as a continual evolution. There has to be a point where ever increasing data rates are just self defeating – unless you really think people will be watching HDTV on the move? If not then ultra-high bandwidth services should come into the home on DSL/Cable or satellite and then be distributed – which could be with LTE, WLAN, UWB (ultra wide band)... No, I think each new air interface and mobile bandwidth just gets that bit closer to 4G but never reaches it – so LTE+ would be 3.99G! 4G goes beyond one technology – after all users don't give a hoot about the access technology, only what they can do with it. Convergence is the first serious candidate for a 4G tag but that can only be justified if the services are somehow better than 3.9/3.5 or 3G. The next chapter looks at mobile services and takes up this question.

References

[1]   Van Veen, N. "Breaking the Mobile Internet's Low Adoption Spell", Forrester (December 2005).
[2]   *Daily Telegraph* 6 Feb. 2008 Business News.
[3]   OMA – Open Mobile Alliance www.openmobilealliance.org/
[4]   *Daily Telegraph* 6 Feb. 2008 Business News.
[5]   Source: Analysis Research, 2007.
[6]   *IP for 3G: Networking Technologies for Mobile Communications*, Wisely et al., John Wiley & Sons Ltd, 2002, ISBN 0471486973, Chapter 5 on Mobility written by Phil "where's the tea" Eardley.
[7]   "SIP-based IEEE802.21 media independent handover — a BT Intel collaboration" K.N. Choong, V.S. Kesavan, S.L. Ng, F. de Carvalho, A.L.Y. Low and C. Maciocco in the *BTTJ*, 2007, vol. 25, no. 2, pp. 219–230, Springer Verlag.
[8]   www.ietf.org/html.charters/mip4-charter.html.
[9]   *IP for 3G: Networking Technologies for Mobile Communications*, Wisely et al., John Wiley & Sons Ltd, 2002, ISBN 0471486973, Chapter 5 on Mobility written by Phil "where's the tea" Eardley.
[10]   "SIP-based IEEE802. 21 media independent handover — a BT Intel collaboration" K.N. Choong, V.S. Kesavan, S. L. Ng, F. de Carvalho, A. L. Y. Low and C. Maciocco in the *BTTJ,* 2007, vol. 25, no. 2, pp. 219–230, Springer Verlag.

More to Explore

Key IMS references

Book *The 3GPP IP Multimedia Subsystem* – Camarillo – John Wiley & Sons Ltd,
2004 – ISBN 0 470 87156 3 – Best of the IMS books – bit detailed but
technically sound and well-written.
IMS News http://www.imsresearch.com/
The following IMS standards are available from 3GPP www.3gpp.org

- **Architecture:** TS 23.002 (3GPP Architecture), TS 23.221 (Architectural
  Principles).

- **IMS Session Control:** TS 22.228 (IMS stage 1), **TS 23.228 (IMS stage 2)**, TS
  24.228 (Rel-5 SIP Signaling Flows), TS 24.229 (IMS Call Control).

- **IMS Services:** TS 23.141, 24.141, 33.141 (Presence), TS 22.340, 24.247
  (Messaging), TS 24.147 (Conferencing), TR 23.979 (PoC).

- **IMS Interworking to CS/PSTN:** TS 29.163.

- **IMS Interworking to SIP Clients in the Internet:** TS 29.162, TR 29.962.

- **IMS Charging:** TS 23.815 (Charging/Architecture), TS 32.200 (Charging
  Principles), TS 32.225, TS 32.235 (IMS Charging Data).

- **IMS Interfaces to Subscriber Database:** TS 29.328, 23.329 (Sh-interface), TS
  29.228, 29.229 (Cx-, Dx-interfaces).

- **Security Mechanisms:** TS 33.102 (Security architecture), TS 33.203 (Access
  Security), TS 33.210 (Network Security), TS 31.103 (ISIM), TS 33.106,
  33.107, 33.108 (Lawful Interception).

- **IMS QoS + SBLP:** TS 23.107, 23.207 (QoS Concept/Architecture), TS
  29.207, 29.208, 29.209 (Go, QoS Signaling Flows, Gq).

- **End-to-end QoS:** TR 23.802 (Rel-7 study, draft only).

- **Evolution of Policy Control and Charging:** TR 23.803 (Rel-7 study, draft only).

- **FBC:** TS 23.125 (stage 2), TS 29.210 (Gx), TS 29.211 (Rx).

- **GPRS:** TS 23.060 (stage 2), TS 24.008 (session management), TS 29.061 (Gi).

- **WLAN Access:** TS 23.234 (stage 2), TS 24.234, TS 29.161 (Wi).

- **Fixed Broadband Access:** TR 24.819 (Rel-7 study, draft only).

SIP

Online tutorial – www.wisely.org/consult/SIP (author's site)
http://www.sipforum.org – links to suppliers, standards, documents etc.
http://www.ietf.org internet standards and drafts
Current SIP draft is RFC 3261

SIP Working Groups:
MMUSIC
SIP
SIMPLE
SIPPING
Excellent SIP site http://www.cs.columbia.edu/sip/

VCC
3GPP TS 23.206 V1.1.0 (2006-07)
Technical Specification
Technical Specification Group Services and System Aspects;
Voice Call Continuity between CS and IMS; Stage 2 (Release 7)

UMA
UMA Specifications
Available from: http://www.umatechnology.org/specifications/index.htm
– R 1.0.2 (UMA Stage 1, 2, 3)
– R 1.0.3 (UMA Stage 1, 2, 3)
– R 1.0.4 (UMA Stage 1, 2, 3)
BT Fusion news http://www.ovum.com/news/euronews.asp?id=6631

Media Independent Handover
802.21 and current standard http://www.ieee802.org/21/
Tutorial – http://www.ieee802.org/21/Tutorials/802%2021-IEEE-Tutorial.ppt
See also Ref. 8 above.

NGN
BT 21C http://web21c.bt.com/
ETSI TiSPAN http://www.etsi.org/tispan/– good for NGN specifications and presentations.

Fixed-Mobile Convergence Alliance
FMC standards body with product specifications http://www.thefmca.com/

Chapter 7: Mobile Services – The Final Chapter

7.1 Introduction

There is no doubt that the first "killer application" in telecommunications was voice, and after more than a hundred years, whilst the fixed network is now dominated by the Internet, mobile revenues are still largely based on voice. Both fixed ("free weekend calls") and mobile ("200 min anytime for £20/month") telco services are still marketed around voice. Users were happy to pay extra – the mobile premium – over and above the cost of a fixed phone for the advantage of having a personal device that went almost anywhere with them. The personalised aspect of mobiles is very important – users really want their own style of handset, ringtone, address book, music, pictures and so on. It is reckoned that to launch a new mobile service requires at least 10 handset types to accommodate the full gamut of users. The other "killer application" of mobile has been SMS – txt 2 U. The fact that SMS was a lucky accident (I can't spell serendipitous) is not always appreciated – it was originally intended for GSM engineers to signal to each other and runs on the signalling channel of GSM. That SMS was a massive hit and that video telephony has been a giant flop should tell you something about how good the mobile (and indeed the whole telco industry) has been at designing, or at least anticipating, services for customers.

It contrasts nicely with our old friend the Internet service model. In the Internet you have a dumb network that delivers packets from edge servers – containing anything from music, games, Sanskrit or whatever – and general purpose end terminals. In the telco model the network elements, terminals and

IP for 4G   David Wisely
© 2009 John Wiley & Sons, Ltd.

service platform are all tightly coupled and interdependent. Telco services are planned in standards, require new handsets, network upgrades and generally happen infrequently. In the Internet services are launched all over the world every day – many fail but that doesn't matter as the barriers to entry are low and some go on to be very successful (a sort of evolutionary approach?).

In this final chapter (which I must acknowledge is based on an article by myself and the legendary Richard "Den" Dennis (who headed up BT's mobility Research Centre for many years before disappearing to Africa this year) originally published in *BT Technology Journal*, vol. 25, no. 2, April 2007) we will look at some exemplar mobile services – successes and failures. We will then try and draw a final conclusion about 4G – based on the technology of the earlier chapters, together with more commercial considerations. The debate about mobile services in the industry has been evolving rapidly over the past few years. VoIP has dropped off the radar, mobile TV has been trialled and finally, Internet on mobile devices – in the form of Mobile Broadband – has been launched. We will look at all of these and I will argue that a single metric – cost/Mbyte – allows a fair comparison between services and neatly explains the coming battle with the fixed Internet.

7.2 Beyond Voice and SMS

Arguably the only "successful" non voice service on mobile phones has, to date, been SMS. The key question that seems to bedevil the mobile industry today is how to get consumers to move beyond voice and SMS. Until recently the figures were not encouraging for operators — in 2005 voice still represented 80–90% of mobile traffic by value and 90% of non voice revenue in 2005 came from SMS (Ref. 1). Figure 7.1 shows that mobile data has risen slowly in the last few years.

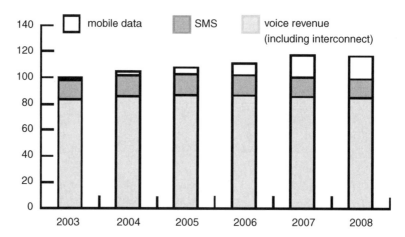

Figure 7-1. Generic European mobile operator retail revenue growth profile. [Source: Adventis].Courtesy of BTTJ. (Source: BTTJ. Reproduced by permission of © British Telecommunications plc.)

Service	2004, %	2005, %
Music	16	57
Browsing	28	35
E-mail	20	26
Mobile TV	N/A	26
Ringtones	16	26
Gaming	20	17
IM/MMS	12	17
Video-streaming	28	17
Videotelephony	4	17

Table 7-1. Responses of the mobile operators to the question: 'Which services do you think will be key in generating revenue on 3G?' Forrester Research Inc (Ref. 4).

Operators have added overall capacity with 3G and now with high-speed downlink packet access (HSPA) (see Chapter 4), but the net result of this extra capacity – combined with massively increased competition from new (e.g. Hutchinson [Ref. 2]) and virtual (e.g. Tesco Mobile) players – has simply acted to reduce mobile voice and SMS prices (see Figure 7.2). When 3G was launched operators had very high expectations that it would be used for a whole range of non voice/SMS services. Table 7.1 (Ref. 3) shows the

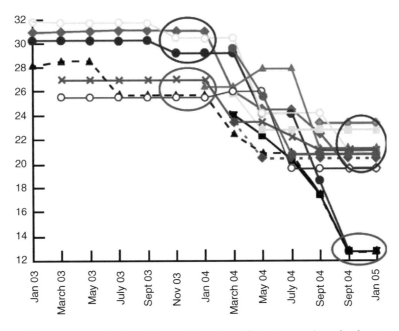

Figure 7-2.   Relative price levels for mobile voice calls in Norway/Sweden from January 2003 to January 2005. Courtesy of the BTTJ. (Source: BTTJ. Reproduced by permission of © British Telecommunications plc.)

expectations of data services that mobile operators were contemplating in 2004/5 – Forrester's analysis of this is entitled 'The Mobile Industry is Littered with Broken Dreams' which very aptly sums up what has happened since.

If you recall back to the dark days of Chapter 2 (when we were talking about UMTS standardisation) the mobile industry's view of services for 3G was originally based on B-ISDN, ATM and video telephony. It then hastily added the IMS – which was going to offer VoIP and video telephony with a range of IN-like services. When 3G launched it was all about "special" content (e.g. football highlights and music in "walled gardens"). Since launch there has been a phase of experimentation – with picture messaging, gaming, music and so on as described above. I think we are now moving into an "end-game" of mobile Internet which we will look at in the final section of this chapter. However, it is time to unpick some of the services in the table and look at the winners and losers and try to understand what lies behind these services.

7.3 Hard Lessons – MMS and WAP

Multimedia messaging service (MMS) – picture messaging in which users take photos with camera phones and send them to other mobile phones or e-mail addresses – looked like a definite winner after the run-away success of SMS. In 2002 when asked "Which services will be the most important for generating revenue during the first three years of commercial 3G services?", mobile operators ranked MMS as the most important (4.2 out of 5 [Ref. 5]). Certainly mobile users are taking pictures – with a full 40% of users having an MMS-capable phone with built-in camera (Ref. 6). Even the GSM Association (basically the GSM operators worldwide) conducted a survey (Ref. 7) in which it claimed that ". . . MMS has become a popular service among mobile phone users worldwide" and that 40% of respondents regard it as "indispensable". Sadly the usage figures paint a rather different picture (see Figure 7.3) with only about 4% of users sending one or more MMS messages per week. What is the reason for this apparent failure? Firstly, there were clear technical limitations with MMS – it was said to have taken over 10 key press and selection operations to send an MMS on a phone by a leading manufacturer. Settings for MMS were also notoriously difficult, with users expected to navigate difficult menu structures to set up MMS. Even OTA (over the air) setting of phones – using SMS – has major limitations. If the user was able to set up MMS, the actual delivery of the messages was poor (as low as 60%) and those that arrived at their destination were often unable to be viewed due to incompatibilities between network implementations and handset issues. The service was also launched too early – before a critical mass of handsets was in use. This led to the classic 'videophone' scenario of having bought a shiny new MMS phone, having to contact all the people that you know and ask them if they have an MMS-capable phone to receive it. The prices for MMS were also set too high – about five times those of SMS which is probably close to the bandwidth premium needed for

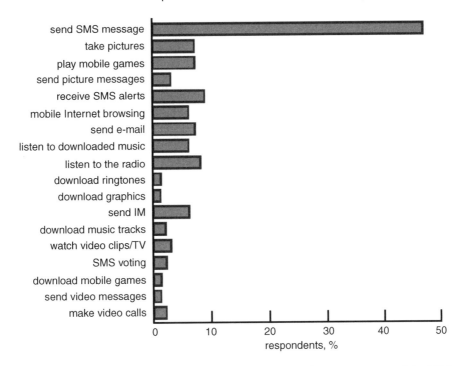

Figure 7-3. Daily usage of common mobile services in the UK. Courtesy of the BTTJ after Forrester (Ref. 3). (Source: BTTJ. Reproduced by permission of © British Telecommunications plc.)

MMS – but it has been consistently shown that users are not prepared to pay the same per-MB prices for all services; instead users are willing to pay in accord with perceived utility (Ref. 8). Thus in a recent survey they claimed to see video telephony as a useful service but were prepared to pay only the same for a video call as a voice call, despite it taking up as much network capacity as 10 voice calls. Finally, users quickly cottoned on to other ways of using pictures taken on a mobile – such as downloading to a PC and sending as an e-mail (almost 100% reliable and free), or printing directly from certain phones or even removing memory cards and printing at supermarkets, etc.

The story of "The Mobile Internet" is worryingly familiar. Initially 3G operators promised "mobile Internet – any time, any place, anywhere". The problems started with wireless application protocol (WAP) which was introduced to offer the mobile Internet on 2G devices. WAP used a modified version of HTML with a simple menu-based interface. Initially this was plagued with the same problems as MMS –technically it did not work very reliably with incompatible browsers that crashed due to lack of standardisation. WAP also initially required a circuit connection – with a per minute cost whether data was being transferred or not, which proved very expensive, and led to very long delays for pages to load. As the general packet radio service (GPRS) was introduced, it was possible to offer users pay per MB and pseudo 'always-on' capability (each transfer session needs a long set-up time in GPRS and 3G due to the non IP nature of current mobile networks) – WiMAX and LTE are designed to overcome these limitations. The problem was that the content providers had largely become disillusioned by WAP and had stopped trying to rewrite their HTML sites in WAP. HTML browsers on mobile phones, however, typically render sites not specifically written for mobiles very badly — resulting in a poor user experience. There are said to be 1000 different mobile phone browsers – all capable of rendering web pages differently. Operators responded to these challenges with a 'walled-garden' approach, with users confined to operator approved or operator run sites with limited content that was specifically tailored to the small screens and limited interaction of typical phones. A good example of this approach was Hutchinson 3 in the early days of 3G (2002–2005), offering music and ringtone downloads, and football highlights. Others (such as O2) invested in the Japanese i-mode technology – which used yet another version of HTML to code sites for mobile terminals – to ensure they worked well.

In the end the operators realised that usage of these services was not growing and opened up their walled gardens to allow users to browse the wider Internet – a good example of this strategy is the T-Mobile 'Web and Walk' tariff, but still revenues are small. Until recently only 4% of users regularly use the mobile Internet (see Figure 7.3).

Recently Vodafone have partnered with Google and Hutchinson 3 with both Skype and Google, to try to overcome this resistance. New handsets and tariffs – such as the 3 X-series [Ref. 9] and T-Mobile's 'Web and Walk' flat fee – are further initiatives by the mobile operators to boost mobile Internet revenues. Part of the problem, however, is that simple information, such as traffic updates and ringtones, is available from other sources, such as SMS, and news/travel is available everywhere. For complex services, such as photo uploading or ordering things, users preferred the PC as it offers a much improved interface experience. We can point to the fallacy that just because there are 3B mobile phones and 800M PCs in the world it does not mean that the Internet is going to change to suit the phones. Worldwide only 7% (Ref. 10) use 3G and of those, they spend 90% of their time using voice and text. When you want to book a holiday or catch up on missed TV or write your thesis you get out your laptop not your phone. PCs have 1000× the memory, 1000× the processing power, 10× the graphics resolution of a phone. The Internet is so engrained that it will never be eclipsed by the mobile Industry. It is also useful to make a distinction between the Mobile Internet – the operator's view of what the Internet should look like on a mobile – and Mobile Broadband – which is a USB dongle for your laptop and connectivity that looks like fixed Broadband. We will consider Mobile Broadband in the final section of the chapter.

Operators have recently moved their expectations from the mobile Internet to music downloads – the subject of the next section – and Mobile Broadband which we will look at in the final section of this chapter.

7.4 Innovation – Mobile Music

Mobile music includes not only downloaded song tracks but also streamed music, ringtones and ringback tones. In terms of mobile services, mobile music is the next largest revenue generating service type after voice and SMS. Figure 7.4 shows that it is expected that the global market for mobile music will reach $30bn per annum by 2010. Ringtones first appeared in 1998 and have progressed from monophonic to polyphonic and, more recently, to 'real tones' – i.e. real pieces of music (also called "true tones" or "master tones"). Driven by the user desire for personalisation and developing into a fashion genre in their own right, ringtones have been very profitable for mobile operators. The share of revenue retained by the operator varies from about 9% in i-mode to 70% in other networks (Ref. 11). However, in mature markets the demand and revenue for ringtones is starting to fall and many commentators see the overall ringtone revenue declining slowly over the next few years. Ringback tones are pieces of music or audio clips that callers hear instead of the normal ringing tone when they dial a specific number. Originally created by Korean operator SKT, they have achieved a high penetration in Asia/Pacific areas, but have yet to make much impact in Europe. Telefonica of Spain has, however, attracted 2 million customers with a low pricing of €1 in the first months and €0.5 in subsequent months – 60% of this revenue is kept by Telefonica (Ref. 12). The real potential for increasing mobile revenue is with full-track downloads – users with music-capable phones downloading paid-for music tracks over cellular networks. There is no doubt that the potential market is huge. However, for the mobile operators to make inroads into this potential income there are several barriers – technical and commercial – that need to be overcome.

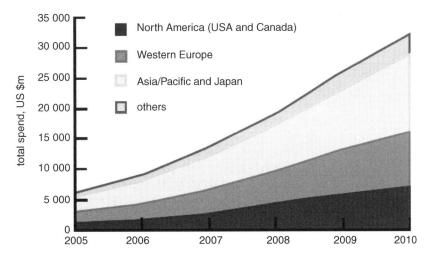

Figure 7-4. Mobile music market – total spending by end users worldwide. [Source: Gartner Dataquest [Ref. 8 October 2006] Courtesy of the BTTJ. (Source: BTTJ. Reproduced by permission of © British Telecommunications plc.)

Firstly, users need to have music-capable handsets – Ovum (Ref. 13) predicts that sales of phones optimised for music will rise from 4 million in 2005 to 71 million in 2010. In addition 378 million music-enabled smart phones and 450 million basic phones with music capability will be sold. This can be contrasted with the dedicated music player market of 135 million in 2005. The first music-optimised phones included the Motorola ROKR E1 (including i-Tunes software) and Sony Walkman phones. Recent developments (such as the Nokia N91) have seen larger memory (4 GB) and better sound quality. Secondly, the issue of digital rights management needs to be addressed. Currently the music download business is suffering from a mixture of incompatible standards and overly prescriptive usage rights. For example, music purchased from the i-Tunes store can only be played on Apple iPods or PCs running i-Tunes. Microsoft's Zune music system is incompatible with Apple's and music purchased with it cannot be played on iPods. Clearly if mobile operators expect users to buy music on their phones, they will need to ensure that those tracks can then be transferred to PCs and other music players consistently. One of the most successful music download services is run by Sprint in the USA. Just seven months after launch the service has reached its 3 millionth download. This is in spite of the heavy premium price ($2.50 compared to $0.99 for the i-Tunes store). Sprint has a total catalogue of 400 000 songs allowing users a 30 sec preview before buying. Sprint has also arranged a number of pre-releases with specific music companies to offer users exclusive access to content, and offered free content such as a Bon Jovi music video. A slightly different model is used by NTT DoCoMo that enables users to download an unlimited number of tunes for a flat fee of $10.90 a month, and to transfer them to another handset for an extra $5 a month.

The major challenge, however, is probably just the sheer number of sources of music – from ripped CDs to illegal downloads to free radio stations. To date, the music industry has conspicuously failed to prevent illegal downloading and ripping to the extent that only an estimated 5% of downloaded music is thought to have been paid for. Overall, however, mobile music looks to be a promising area of data-service growth for mobile operators over the next five years or so. Users have become used to portable music players and buying tracks online. The mobile phone, with its SIM-based security and billing system, is well placed to exploit this. In addition, the general trend of larger and cheaper memories and more capable handsets means that performance comparable to that of today's mid-range dedicated music players is within the reach of mobile handsets over the short term. If mobile operators can get access to the right content, package and bundle it attractively with exclusive content, then there seems to be a genuine belief in the industry that mobile music will exceed the current ambitious growth predictions.

7.5 Broadcast Services

Mobile broadcast (or multicast) is a relatively recent service innovation for mobile service providers. Things are technically very complicated in the broadcast services space with three trials in the UK alone during 2006 – with Virgin, O2 and Orange. Each trial used a different technology and, currently, there are ten technological solutions for mobile broadcast, in addition to the (brute-force) solution of setting up one circuit connection on 3G for each customer (as Orange did in their trial). Using one channel per connection is very resource intensive and would, if there is ever mass take-up of the service, result in significant congestion even with new capacity from 3G and HSDPA. Broadcast services are very important in the wider convergence market – with triple- and quadruple-play operators trying to move beyond offering a single bill for fixed phone, mobile phone, broadband and pay TV. The market is awash with new offerings of TV over broadband, and TV over mobile is predicted to be the next 'big thing' in this sector. In this section we look at the three technologies most relevant to the UK and Europe (see Poulbere [Ref. 14] for a full description of all the available technologies).

7.5.1 DVB-H – Digital Video Broadcast-Handheld

DVB-H is a broadcast technology that can be used either from a terrestrial infrastructure or from satellites. Figure 7.5 shows the terrestrial operation. In this section we will only look at terrestrial DVB-H as this is now operating in

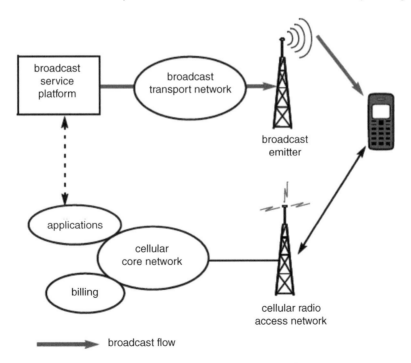

Figure 7-5.   Terrestrial broadcast solution (Courtesy of the BTTJ). (Source: BTTJ. Reproduced by permission of © British Telecommunications plc.)

Europe and formed the basis of a trial in the UK by O2 in Oxford in 2006 (Ref. 15). DVB-H uses either the UHF band (470–862 MHz) or L band (1.4–1.5 GHz) with a channel bandwidth of 5–8 MHz giving a throughput of 3–12 Mbit/s and sufficient for 15–25 channels. Major technical features of DVB-H include time division multiplexing to allow power saving and support for IP-based traffic (such as web content). One of the major issues for DVB-H is clearly the shortage of broadcast spectrum. Only with the analogue TV switch off (planned for 2008–2012 in the UK) will sufficient spectrum become available. Currently the only commercial services in Europe are from Mediaset/TIM and 3 in Italy and Digita in Finland. There are plenty of receivers available for DVB-H including the Nokia N92 and Samsung P920, and the technology is backed by all the leading handset manufacturers. Aside from spectrum issues, the other major hurdle for DVB-H to overcome is the high potential cost of network roll-out — Ovum (Ref. 16) calculates that to cover 90% of the UK population would require 2300 base sites for the UHF band and more in the L band. Studies have shown that neither existing broadcast nor cellular sites are suitable and therefore many new sites would be required.

7.5.2 DAB-IP – Digital Audio Broadcast-IP

DAB-IP is an evolution of the original DAB standard that is familiar from digital radio broadcasts that have been available in the UK and Europe for several years. DAB-IP adds IP delivery mechanisms and supports Windows Media codecs and digital rights management (DRM). DAB-IP can operate in either the VHF band III (174 to 230 MHz) or L band. It uses a 1.5 MHz channel for a throughput of 1.1 Mbit/s, sufficient for two to four channels. This technology – in conjunction with the BT Movio content delivery platform – is used by Virgin mobile to offer a limited commercial service in the UK (Ref. 17) with common TV channels available. DAB-IP can potentially reuse existing DAB infrastructure when there is sufficient capacity and can easily be integrated with DVB-H solutions. However, currently no major handset vendor is committed to the technology and the narrowband DAB channels have limited capacity.

7.5.3 MBMS – Multimedia Broadcast and Multicast Services

MBMS is the cellular industry's technical solution to broadcast and multicast to mobile devices. MBMS uses spare capacity on the existing 3G spectrum and so requires no new spectrum. It does, however, require major upgrades of both the access as well as the core 3G network – with a new element, the broadcast multicast service centre (BM-SC) needed in the core. MBMS is included in UMTS Release 6 and first implementations were expected by the end of 2007 with commercial trials in 2008. Figure 7.6 shows the architecture of the solution.

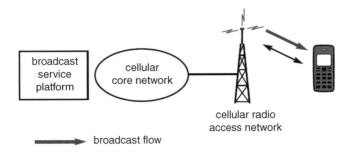

Figure 7-6. MBMS solution (Courtesy of the BTTJ). (Source: BTTJ. Reproduced by permission of © British Telecommunications plc.)

MBMS is restricted by the limited capacity available on 3G – operators can partition how much capacity is used for broadcast and how much for unicast applications, such as voice, in a dynamic way. MBMS has the key advantages that no new spectrum nor base sites are required.

7.6 A New Paradigm – 1000 Killer Applications

It is possible to go further than voice and SMS and consider services that use these underlying mechanisms. Examples might include chat lines, recorded information, reverse SMS billing, ticket purchase and many, many more. Indeed Ahonen recently claimed to have detailed 1000 'real mobile services' (Ref. 18). In this section we look at some of the diversity of such services – in an attempt to understand how these services might evolve and how operators can encourage and promote such services in order to drive traffic and revenue growth. A Belgian study by the Catholic University of Louvain of 2500 Belgian teenagers revealed that young people are getting less sleep because they send text messages at night from bed. 20% said they are regularly awakened by incoming SMS (Ref. 19). SMS is also the key to dating for the younger generation – no more corny pick-up lines – now a boy must send a witty and amusing SMS if he is to ask a young lady out. This allows the girl time to consider, showing it to several of her girlfriends and makes rejection easier for both parties. If a relationship starts then a very minimum of an SMS a day is expected (Ref. 20). A recent study of phone use by Motorola (Ref. 21) describes a number of 'tricks' to get dates with boys/girls. One technique is to take a picture with the camera on your phone and offer to send it to the subject – if only they will tell you their number! In Dubai it is reported that young men are buying phones with a pre-paid SIM and handing them to attractive girls.

Mobile TV is another strong growth area for mobile operators – a recent report (Ref. 22) predicts mobile TV will grow by 50% a year for the next five years. One of the key growth areas to date have been mobisodes – special episodes of popular shows and series that have been edited for mobile terminals and shortened to a few minutes for viewing at a bus stop or in a coffee queue, for example. Last year there were over 250 000 live streams on Hutchinson 3 during the first three days of Big Brother. Scott Taylor, 3's head of content and services, said: "The results so far are outstanding. The intrigue in this year's show is definitely driving interest but the substantial upswing we're experiencing is because customers are hooked on the intimacy and interactivity of the 3G experience". Wider afield, ABC are planning a series of two-minute mobisodes of the Lost series – using actors and writers from the show and Cingular Wireless have produced mobisodes of Entourage with writers and actors from the series – the cost is $5 a month.

Payment services have been available in the Asia/Pacific areas for several years now and are starting to appear in Europe with SMS payments for car parks, for instance, being a typical application. Moving beyond this is a development from Nokia – called near field communication (NFC) that allows users to pay for goods and services simply by waving their phone over a reader – much in the same way that "Oyster" payment cards are currently used on London Transport. NFC is a derivative of RF-ID (radio-frequency ID tags that are used to keep tabs on goods and for security cards) with the ability to read and write small amounts of data. In the Asia/Pacific areas many train tickets are now purchased with similar systems – in Japan, for instance, the Suica system uses the phone as an RFID tag on most underground and train stations in Tokyo, with software on the phone deducting the fare from your account. These services are really important for mobile operators for many reasons. Firstly, they can be substitutional services – meaning that payment services and operators already exist and that the mobile operators would then be capturing this value. We have seen how hard it is to devise and launch new mobile services – so adapting existing ones for mobile is, in many ways, a much more promising line of attack. Secondly, IP and payment services are a natural extension of the mobile concept of "owning the customer". In the Internet it is almost impossible to "own the customer" for anything other than a single service. Your ISP knows about your access but nothing about your book-buying or music listening and has no role whatsoever to play in these. By contrast, if you buy music from a mobile operator's portal the payment can be part of your bill or taken from your Pay as You Go balance. The SIM card is the technology on which this control has been based. In fact the SIM card has moved on to be a complete smart card with large memory (512Kbyte +), secure processing environment for applications and an API to the phone operating systems. Operators are keen to exploit these as the secure storage area for NFC keys and algorithms. There is a tension here with the mobile phone vendors who favour a more open system – with the secure credentials stored outside of the SIM on an independent tamper proof chip. This neatly illustrates the tussle between the operators and vendors for this lucrative authentication, credential and authorisation space. But they are not the only players. In my wallet I have (apart from very little money – big spending wife and two kids to support): An Oyster card (London Transport smart card); two credit cards with smart chips on, a BT pass with smart chip on for building access. In another pocket I have a mobile phone with a SIM card. Why do I need all these cards which are essentially smart cards of one type or another?

Technically there is not much reason – as we have seen with mobile phones being used (through NFC) as Oyster cards and for payment of parking. Phones are also being trialled (with NFC) to pay for small purchases such as coffee and newspapers.[1] No, the reason that I have so many cards is commercial. If the banks and other organisations let the mobile operators take over these functions it would cost money (because the operators would want a cut) and they would lose ownership of the customer. At the moment I am unlikely to change my credit cards or suppliers but if they were on my phone and I could change supplier at the touch of an advert, say, then the mobile operator would start to play a much more important brokering role in the process. These are the kinds of tussle we can expect as the different players vie for control of what is fast becoming a converged battleground.

Of course it is easy to get carried away with thinking about high bandwidth 3G services when 93% of the world's GSM connections are only 2G. One of the more surprising 2G services that is gaining ground in countries such as Kenya and the Philippines is mobile remittance. Safaricom's M-PESA in Kenya had, in its first 11 months acquired 1.6 million subscribers with about £75 million worth of transactions made. Typically the service might be used by a migrant worker without access to banking facilities. They would then buy (Pay as You Go) top up for their mobile phone. This can then be used to send money back home – where families can receive the money sent in cash at a local merchant. In the Philippines money can be transferred to Globe's GCash – an e-wallet system – and money redeemed in cash at various outlets or sent on in the form of an SMS.

There are literally thousands of these services – relying on basic communications services – the innovation being driven by a range of different factors such as markets, technologies, fashions and user ingenuity. The key for operators of all types is to successfully tap this growth, while at the same time controlling enough of the value chain to make a significant return.

7.7 Terminals

Terminals are the very life blood of mobile services (steady on Wisely this isn't a novel you know) – but, more prosaically, new services have always gone hand in hand with new terminals (think mobile Internet, Fusion, Mobile Broadband). The reason for that is pretty much that they are dedicated hardware and software with limited upgrade capabilities and an estimated service life of about 18 months. Compare that to a Windows[2] laptop – with a service life of four years (five in BT) – it is regularly updated, has 100 times the processing power

---

[1] And we have had a few jokes about buying 10 packets of crisps by leaning on the vending machine!

[2] It is completely untrue that I am on a bung from Microsoft – I think their products are superb – nothing touches Windows. Windows is quite simply the best operating system in the universe – the Vogan invasion only failed because of a bug in the video sharing application built into their Black Panther OS.

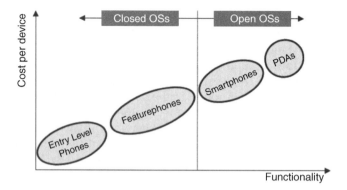

Figure 7-7. Mobile phone OS types. (*Source:* BT. Reproduced by permission of © British Telecommunications plc.)

and 1000 times the memory of a phone. My laptop can cope will all the new services launched over the last few years – and I can upgrade all the components. On mobiles this tight coupling between services and terminals means that the two tend to go hand in hand – you need a camera phone for MMS and video-telephony and a music player if you are downloading music. As I have said before it is an industry truth that you need 10 phones to launch a new service.

There are different categories of mobile devices (Figure 7.7) – with entry level phones having completely fixed and closed (no downloading games, no adding features) OS systems. Feature phones have more – well features – a camera, possible Java for games but, again have closed OS systems with no third party applications. They have very limited memory and processing and offer a limited range of dedicated services.

Smartphones – such as those running Symbian and Windows Mobile OS offer both a Java and a native OS interface for developers (Figure 7.8). The Java Virtual Machine offers developers – using Java 2 Mobile Edition (J2ME) – access to a subset of the OS features. Native applications are able to exploit a well-known API and have more access to lower layer functions.

Even a smart phone is very limited compared to a PC – the screen is much smaller, the processing power much reduced and the memory limited. One of the key questions for mobile services going forward is whether this will change to any significant extent? What if new displays were available that could be unrolled to 10″ × 8″ say or if retinal scanning – where light is written directly to the retina –takes off? Screen technology is continuing apace – IO2 Technology have launched a display to project video images onto thin air – displaying video without the need for a screen. Polymer Vision, a spin-out from Philips, has developed a rolling display giving medium quality with four levels of grey. Many other developments are in the pipeline – although it is fair to say some have been in the pipeline a long time.

Until these developments happen the small screen of mobiles will remain a bugbear of the Mobile Internet. There have been many attempts to convert or re-write Internet pages to fit on small screens – WAP, i-Mode, walled gardens of

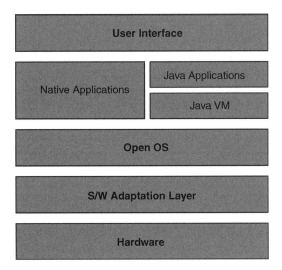

Figure 7-8.   Typical smart phone OS stack – e.g. Symbian. (*Source:* BT. Reproduced by permission of © British Telecommunications plc.)

special content. In the end they have all failed because, I think, they were very limited compared to the Internet experience on a PC (especially as broadband has become ubiquitous).

Memory is getting cheaper – I bought a 4Gbyte USB stick for £24 yesterday at the supermarket (Tescos). The memory chips probably cost less than £5. So there is little barrier to very large memory chips appearing in phones – even up to 1Tbyte is talked about in a few years time. Why on earth would you want 1Tbyte storage in your phone? Caching – on a smallish screen a hour of video takes up about 200–300Mbyte – even a full DVD is only 5Gbyte – so you could store 5000 hours of video for a mobile or 200 full DVDs. Music is even better – each album compresses to 30–50Mbyte – that's 20 000 albums to a Tbyte! At a stroke all that lengthy downloading is avoided – all that needs to come over the network are keys to unlock the media. Maybe the cache is updated when the network is uncontested (at night) or when the terminal is near a Femtocell or WLAN hot spot to put in the latest movies. Some of these ideas are already being tried out. Terminals will continue to get more powerful but whether they can ever converge with the PC/broadband Terminals experience is a moot point – Google Earth, Second Life – that's a lot of processing and graphics.

7.8 Money Makes the World Go Around

There is only one reason mobile operators introduce services – to make money. The question is what is a fair price for a mobile service? Actually that is not the question – because the price that operators charge for services has nothing much to do with cost and everything to do with what the customer is willing to pay – subject to competition and regulation. There has always been a mobile premium – where the same services cost more on a mobile than on a fixed phone/connection. At its highest this was, 10 times for voice and 100 times for voice roaming (even higher for roaming data – we will come back to this). In 2006 there were three trials of TV over mobile using different technologies, as we noted earlier. The major conclusion seemed to be that users liked the service but were not prepared to pay much more than about £7 a month for it. It was used surprisingly often indoors and the most popular location for viewing was the bathroom (Ref. 23) which makes you wonder why people don't just install a TV in there? Video calling, similarly, is liked by the public in studies but users are unwilling to pay much more for a video call than for a voice call. SMS, however, is a very popular service – 2.2 billion messages sent on Valentine's Day 2008 in North America alone.

All of which, in a rather rambling way, brings us to the concept of ARPU (said R-poo) – Average Revenue Per User. This is an absolutely crucial metric in the mobile industry – watched by analysts and investors very closely. ARPU is important because the revenue of your network is: number of users*ARPU. So the only way to grow is to recruit new users or improve the ARPU. Of course different groups have different ARPUs – Pay as You Go ARPU is lower than for people on contract ARPU. Corporate ARPU is higher than consumer. The problem with ARPU is that, in mature (saturated) markets, improving it is the only way to grow the business[3] So growing ARPU is good but to do that the mobile operators either need to put up prices for existing services (e.g. voice) or find new services. As we saw in Figure 7.4 mobile prices have been falling in mature markets due to a combination of competition (new operators and new virtual[4] operators) as well as regulatory pressure. So the squeeze is on mobile operators to come up with new services to drive up ARPU. Figure 7.9 shows a prediction of ARPU made in 2000 (Ref. 24). You can see that the authors (the EU TONIC project) clearly thought that by 2008 voice ARPU would shrink (which it has) but they got it completely wrong on SMS – which has gone on growing – and video/audio streaming and Internet which has still not taken off. What has been accurate is that no new killer application has emerged to challenge voice and SMS and, as the price of these has declined, ARPU has remained constant as users have taken up a mixed basket of new services such as ringtone, MMS and music downloads.

The big question for mobile operators in the future is where is the new ARPU coming from to make up for continuing voice/SMS price reductions? One very useful way to look at this is by plotting the revenue per Mbyte for services versus the required bandwidth (Figure 7.10) – an approach used by Analysis (Ref. 25).

---

[3] Of course profits might rise even if ARPU and numbers were static – if costs fell. Some mobile costs have fallen (e.g. the price of backhaul) but other have risen (cost of spectrum, site acquisition, customer acquisition) – so overall this has been broadly neutral.

[4] Virtual mobile operators – MVNOs – do not have a network themselves but rely on a network operator. The MVNO is usually responsible for marketing and billing and some have innovative pricing for particular market segments (e.g. in the UK Virgin Mobile targets the youth market).

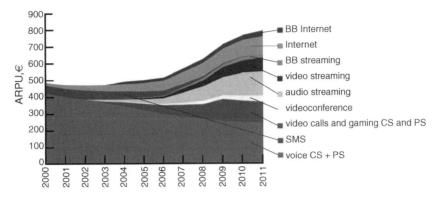

Figure 7-9.   ARPU prediction made in 2000 (EU TONIC project). (Courtesy of the BTTJ). (*Source: BTTJ.* Reproduced by permission of © British Telecommunications plc.)

This interesting approach reveals that most existing mobile services fall on a line of decreasing revenue per MB with increasing required network throughput. This reflects the facts described above whereby users will pay 1.5 times the cost of a voice call to add video but not the 10 times increase in capacity actually required, or £7 a month for mobile TV when the same volume of data, charged as mobile voice, would cost nearer £100 per month. The graph also shows that the cost of supplying a MB of traffic gets much lower on the newer

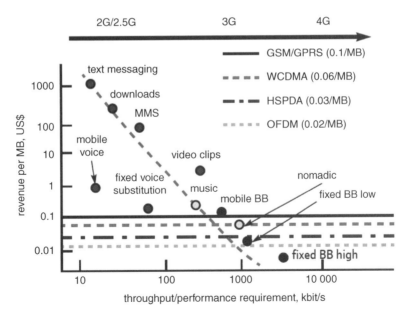

Figure 7-10.   Revenue per MB versus required network throughput. [Source: Analysis] (Courtesy of the BTTJ). (Source: BTTJ. Reproduced by permission of © British Telecommunications plc.)

networks — through HSDPA, LTE and WiMAX and that the profit margin of the operator is represented by the difference of revenue minus cost (less the cost for the content from the copyright owners, if any). It is easy to see why services such as ringtone download, MMS and SMS (all of which require little network capacity) are so important to mobile operators and why they are keen to avoid cannibalisation of these revenues.

Figure 7.10 shows clearly some of the major challenges going forward. Firstly, there is the "holy grail" – services that would, if they existed (!), sit in the upper right of the diagram with both high bandwidth requirements and high revenue capability. This is the region that WLAN operators would like to sit to differentiate themselves from cellular operators. It is also the quadrant that WiMAX and LTE are really targeting – again their differentiator is high bandwidth.

Another challenge that the graph illustrates is that of falling prices – services often begin at the top of the graph and "sink" over time – meaning that the bandwidth stays constant but the price goes down over time. So ringtones and SMS started at the top left and are now sinking towards the bottom left as they are increasingly part of bundles (one UK operator offers free weekend texts – the small print says that if you send more than 3000 your service will be suspended. If I sent 3000 texts in a weekend I would need finger surgery). Voice is also sinking over time and the point on the graph represents an average 10 ppm (pence per minute) – you can get much cheaper deals (3 ppm) if you are a heavy user.

Where does VoIP sit on the graph? According to Analysis (Figure 7.11) this is much more expensive than CS voice in R99 and HSPA networks – not to

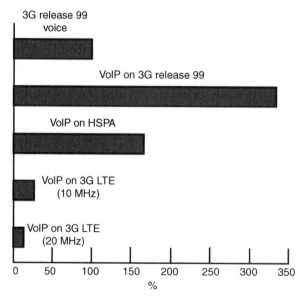

Figure 7-11.   Relative cost of VoIP on mobile networks (Source Analysis) (Courtesy of the BTTJ). (Source: BTTJ. Reproduced by permission of © British Telecommunications plc.)

mention the long delays. Only with LTE and WiMAX does VoIP become a realistic possibility. However, there is some question as to whether LTE and WiMAX will offer standard voice: At a recent conference Vodafone stated that they did not expect LTE to support voice and that the GSM network would continue until 2018 at the earliest (Ref. 26). Certainly LTE may be used exclusively for data services with operators falling back to 3G or even 2G for plain vanilla voice. If the cost predictions of Figure 7.10 are true, however, then it is possible that WiMAX/LTE will finally usher in the era of "free" mobile voice calls. Nothing is for free in this world, of course. You will still have to pay a subscription and it will be limited by a fair use policy but I am convinced it will come. Prices don't fall, year on year, by 10–20% and then suddenly stop just because they are getting near zero. I think the same applies to fixed broadband – for which prices fell 40% from 2004 to 2007. Now you can get "free" broadband in the UK – if you take Sky TV or Talk-Talk's phone package. Not everyone agrees but this is certainly one view of the future.

In Figure 7.10 the horizontal lines represent cost – but price doesn't always have much relation to cost – prices are mostly set to maximise profit in the long term.

The final point about the cost/bandwidth graph is that it has a floor – you can hit rock bottom! The bottom line in business is the cost – not many businesses have survived with a price lower than their cost for long (or a big government subsidy!). The graph's variously dotted line shows that Analysis believe that you can come up with a meaningful figure for the cost of providing 1 Mbyte of data over a mobile network – I think this is fair enough but caution must be attached to the actual numbers as the cost of mobile networks is very complex including such things as:

- Site acquisition cost;

- Site rental and maintenance;

- Backhaul costs;

- Interconnect costs;

- Licence costs;

- Customer acquisition cost (advertising, shops, retention);

- Handset subsidies;

- Fraud.

Some costs are rising (Customer acquisition) and some falling (backhaul). Overall I think the numbers are indicative of what might happen to costs. And that is the point – if the mobile operators give up on trying to sell users fancy ringtones, music and picture messaging – they can always sell them capacity at cost plus 20% (say). There is a killer application that eats data, has thousands of wonderful applications already, a proven business model and billions of addicts – it's called the Internet! I don't mean the Mobile Internet silly – that is about rendering pages on small screens, portals, protected content, high

Figure 7-12.   The author hard at work in the bus cafe on the A12 close to Martlesham (good breakfast). (Source: Author. Reproduced by permission of © Dave Wisely, British Telecommunications plc.)

prices per web page. No I mean connection to laptops with broadband-like packages – 3–5GByte per month, straight connection and a fixed fee – I mean Mobile Broadband.

7.9 Mobile Broadband

Figure 7.12 shows the author writing this book in a bus cafe on the A12 in Suffolk – I find the office very distracting and like to work in cafes. My laptop is connected to HSDPA and I was getting rates up to about 500Kbit/s. These products have been appearing over the past year or so from 3G operators – but in the past the cost has been high (see Table 7.2).

Operator	Package	First year cost	GB/month	£/GB/month
3	Broadband Plus	£250	3	£83
3	Broadband Max	£350	7	£50
O2	Web Max	£398	3	£133
Orange	Business Everywhere	£348	3	£116
T-Mobile	Broadband Lite	£220	1	£220
T-Mobile	web'n'walk Plus	£348	3	£116
T-Mobile	web'n'walk Max	£528	10	£53
Vodafone	Mobile Broadband	£417	3	£139

Table 7-2. Prices of 3G mobile internet summer 2007.

However, there has been a dramatic reduction in price in the UK – mirroring that in Europe. There are now offers from £5/month to £25 for 1–7Gbyte of data with varying lengths of contract and varying costs for the modem. If we take a typical deal – £15/month gets you 3GByte of data. If you used 2Gbyte on average then that works out about 0.75pence per MB – looking back to Figure 7.10 you can see this is close to the estimated cost on HSPA.

Questions have been raised about the sustainability of these products and the impact of network congestion if they continue to sell at a reputed rate of 50 000 a month in the UK alone since these price reductions were made. This is one area where Femtocells could be used to offload some of this data traffic to the DSL backbone. Other options include deploying more CDMA carriers to busy sites (if these are available in existing spectrum), rushing out HSPA release 6 and 7 solutions that improve efficiency through better receivers and longer term solutions such as denser network sites and technologies such as LTE in new spectrum. It is fair to say, however, that only Femtocells offer the prospect of carrying heavy Internet traffic – such as IPTV and file sharing or DVD downloads say.

If Mobile Broadband really takes off then it has the possibility to pull users away from WLAN hot spots and even from fixed broadband. In Austria, since April 2007 Mobile Broadband has been priced at €25 a month compared to fixed at €35 and in Q2 2007 it is reported that 64% of all new broadband users were mobile and 20% of the total broadband base was mobile (Ref. 27). Clearly there will be price pressures on WLAN as well as on fixed broadband.

7.10 Conclusion

If it is not possible to define 4G by means of a technology, spectrum, network or convergence, then maybe it is possible to define a 4G "service set". The problem with coming up with such lists is that, in the past, they have been proved totally incorrect. In the case of 1G many people thought that it would be used for emergencies only and that the voice quality would be so poor that people would quickly find a landline to continue the conversation. It has been said that the PSTN "killer application" was voice and the killer application for 1G or 2G mobile was voice coupled with mobility. Users were prepared to pay a "mobility premium" for a service with which they were already familiar. With 3G the service set was initially described as "any thing, any time, anywhere" and then as the "mobile Internet" – however, these predictions turned out to be wildly optimistic. 3G has been more about cheap voice, which it is technically well suited to provide. Even circuit-based services, other than voice, such as video calling and watching highlights of football and cricket matches, have not generated significant revenues. Currently mobile data accounts for only a small fraction of revenue and, of this revenue, over 90% is generated by SMS.

I think there are three possible futures for the mobile industry:

7.10.1 Domination by the Mobile Operators

In this scenario the mobile operators gain significant power – taking "ownership of the customer" – selling bundles of services and seeking to gain revenues from areas such as mobile payments and mobile TV.

Major features of this scenario would be:

- Aggressive roll out of HSPA phase 2 and then LTE in 2010;

- Mobile Broadband being pushed aggressively;

- Broadcast solutions (e.g. MBMS) that are more efficient than existing solutions for TV/video distribution services;

- Widespread Femtocell deployment to increase capacity at lower cost;

- Partnerships with major Internet players;

- Smartphones with large memories for content storage;

- Payment systems – such as NFC being aggressively pushed and deals done for payment with key commercial players;

- Mobile operators selling bundles of broadband and TV.

This scenario would see the importance of WLANs diminish – in South Africa WLAN hot spot traffic has peaked and is falling following the aggressive roll out of Mobile Broadband. It would see users with more personal value invested in their handsets – which would be used for many payments and authentication functions. Music and video would be increasingly stored on the phones – even if it is played on other devices. There are 637 HSPA devices with an annual growth rate of 150% (Ref. 28) – meaning that it is not impossible that cellular connectivity is built into everything in the way that WLANs have been over the last few years. Mobile operators would not be able to dominate the fixed Internet – simply because the bandwidth and data volumes of application such as HDTV, P2P and IPTV are too high. They would seek, however, to own the customer and to offer either fixed Internet retail themselves (as many of them do today) or to use Femtocells to allow distribution to standard devices. Partnerships with the Internet giants – Google and Microsoft might emerge, as well as ownership or partnership with key content owners (such as music and sports).

7.10.2 Converged World

The second scenario that I think is possible is that a converged world emerges. This is characterised by:

- Multiple access technologies co-existing (WLAN, WiMAX, Cellular etc.);

- Seamless handover technologies;

- Service providers with technologies such as the IMS for service creation;

- More open terminals – with built in WLAN, WiMAX and cellular radios;

- Prices falling for connectivity due to widespread competition;

- More advanced service on fixed DSL – VoIP, IPTV;

- QoS and stronger security on DSL and WLAN.

In this scenario the key players are the service providers – they are the ones that "own the customer" and control: QoS, security, location information. They use this power to deliver content and services – possibly created with a platform such as the IMS – to users over a range of access technologies. The actual access technology could be anything from a free WLAN to cellular 3G – with less value in the connectivity than the service. QoS would be needed on WLANs and DSL to allow VoIP and other real-time services to survive during congestion and justify a premium over current best-effort services. In addition stronger security would be needed on WLANs and DSL – so users could start to make purchases on the account at the service providers rather than using a credit card. Terminals would be more open – with WiMAX and WLAN chip cost falling and being included in most new laptops, PDAs and phones. Technologies such as Media Independent Handover would be needed to reduce power consumption and allow seamless handover.

In this scenario it is not so much the access that matters – that has more or less become ubiquitous and is not providing the bulk of the revenues. After all, Mobile Broadband is being provided at close to cost and is, even in Austria, yielding only about 25% of the non voice ARPU of operators. No, the key players will be the middle men – who provide all the location/QoS/billing/trust etc. – that sit above the connectivity and below the content. Authentication, digital rights and authorisation will be the keys to "owning the customers".

7.10.3 Internet Scenario

Finally there is the Internet scenario. The Internet is just so big that it comes to dominate mobile as well. The Internet business model of clean separation between different functions, end-to-end intelligence and access indifference will come to overtake the mobile world.

This might be characterised by:

- Well-known Internet players launching phones – e.g. Apple and Google;

- Mobile Broadband becoming cheap and ubiquitous;

- WLAN and WiMAX Internet connectivity offering low cost Internet connectivity;

- Mobile operators being forced to partner with the Internet giants;
- The Internet business model being introduced in the mobile world;
- New screen and technology advances offering a better mobile Internet experience;
- The mobile premium disappears;
- Internet giants (Amazon, Google, Microsoft) own the customer.

If mobile (3G, HSPA and LTE) joins DSL and WLAN as just basically a mechanism for connecting to the Internet then the value of (say) buying a book will be split between the connectivity supplier, the search engine that finds the bookshop, the review site, the book seller and the credit card company. If this scenario comes about then the Internet giants will increasingly make their presence felt – as Apple have done with the i-phone. Internet services – such as Skype – will appear as the mobile network becomes more IP friendly (with HSPA and LTE).

Mobile Broadband is a major step on this road – when my group recently tested VoIP clients over Mobile Broadband we found that they mostly worked fine over a number of mobile HSPA networks! Laptops are being given away at my local branch of PC World if you sign up to Mobile Broadband! I think the key stumbling block is the terminals – at the moment your laptop is not really portable or sufficiently low power to use as your voice terminal and your mobile smartphone is just too small to access the Internet properly.

7.10.4 Final Conclusion

I can't predict the future – if I could I would be down the bookies – so you are going to have to weigh up the evidence and make up your own mind. Is IP for 4G the mobile industry solution – LTE? Or is a converged world of "IP over everything and everything over IP"? Or is just the Internet translated into the mobile space using the Internet business model? All three are valid as a definition of 4G. What is certain is that mobility is marching on – WLANs, HSPA, WiMAX are all helping to satisfy the user need to be "mobile" in some sense. At the same time these technologies are blurring the distinction between fixed and mobile – in many cases using the same backhaul technologies such as DSL and Ethernet over fibre. WLANs are growing more like cellular technologies by the day – with enhanced QoS, range, security etc. – and cellular technologies are getting more like WLANs – with pico and Femtocells and the shifting of voice back to GSM. In the end the users "don't give a stuff" about the technology, it is the service and content that attract them. If the Internet going mobile means users get to the content and get the services they are prepared to pay to consume then, for me, that is a very good definition of 4G.

References

[1]   Van Veen, N. "Getting Consumers To Use Mobile Services", Forrester (September 2006).
[2]   3 X-Series — http://www.three.co.uk/xseries/index.omp
[3]   Van Veen, N. "Getting Consumers To Use Mobile Services", Forrester (September 2006).
[4]   Van Veen, N. "Getting Consumers To Use Mobile Services", Forrester (September 2006).
[5]   Van Veen, N. Breaking the Mobile Internet's Low Adoption Spell, Forrester (December 2005).
[6]   Van Veen N. "Getting Consumers To Use Mobile Services", Forrester (September 2006).
[7]   Lonergan, D. "Driving usage of value-added mobile services", (December 2005).
[8]   Wisely D.R. "Cellular mobile – the generation game", *BT Technology Journal*, 25, No. 2, pp. 27–41 (April 2007).
[9]   3 X-Series — http://www.three.co.uk/xseries/index.omp
[10] Alan Hadden, President, GSA, Global Mobile Suppliers Association, Informa Telecoms & Media LTE World Summit, Berlin, 21–23 May 2008.
[11] Pittet, S., McGuire, M. and Ingelbrecht, N. "The Outlook for the Global Mobile Music Market, 2005–2010", Dataquest Insight (November 2006).
[12] Poulbere, V. "Mobile music devices", Ovum (July 2006).
[13] Poulbere, V. "Mobile music devices", Ovum (July 2006).
[14] Poulbere, V. "The fragmented landscape of mobile broadcast solutions", Ovum (August 2006).
[15] 3G DVB-H trial — http://www.3g.co.uk/PR/Sept2005/1948.htm
[16] Poulbere, V. "The fragmented landscape of mobile broadcast solutions", Ovum (August 2006).
[17] 3G DAB-IP trial — http://www.3g.co.uk/PR/Sept2006/3582.htm
[18] Ahonen, T. T. "Striving for Killer Apps in 3G LTE: Areas Showing Most Promise", 3G LTE Conference, The Cafe Royal, London, UK (October 2006).
[19] Textually (January 2004) http://www.textually.org/
[20] Ahonen, T. T., and Moore, A. "Communities Dominate Brands", Futuretext Ltd (2005).
[21] Motorola Generation Here, Global Survey (April 2006).
[22] IMS Research – http://www.imsresearch.com/
[23] BBC – http://news.bbc.co.uk/1/hi/technology/4271474.stm
[24] Cordis (ARPU) – http://cordis.europa.eu/fetch?CALLER=PROJ_IST&ACTION = D&RCN=53630&DOC=1&CAT= PROJ&QUERY=1152669942082.
[25] Heath M., Brydon A., Pow R., Colucci M., and Davies G. "Prospects for the Evolution of 3G and 4G, *Analysis*. (2006)
[26] Mike Walters – Vodafone – comments at the Informa Telecoms & Media LTE World Summit, Berlin, 21–23 May 2008.
[27] "Identifying the Challenges of Competing with Fixed Line Broadband" Arthur D. Little, Ansgar Schlautmann, Informa Telecoms & Media LTE World Summit, Berlin, 21–23 May 2008.
[28] Alan Hadden, President, GSA, Global Mobile Suppliers Association, Informa Telecoms & Media LTE World Summit, Berlin, 21–23 May 2008.

Index

IP for 4G David Wisley
© 2009 John Wiley & Sons, Ltd.